# Key Technologies for the 21st Century

• • •

SCIENTIFIC AMERICAN : A SPECIAL ISSUE

W. H. FREEMAN AND COMPANY
New York

**Library of Congress Cataloging-in-Publication Data**

Key technologies for the 21st century.
p. cm.
On t.p.: Scientific American, a special issue.
Includes bibliographical references and index.
ISBN 0-7167-2948-2 (soft cover)
1. Technological innovations.
I. Scientific American.
T173.8.K49 1996
601.12—dc20 95-49200
CIP



The 24 chapters and the commentaries in this book originally appeared as articles in the September 1995 issue of SCIENTIFIC AMERICAN.

Printed in the United States of America

First printing 1996, RRD

# CONTENTS

## PART III: MEDICINE



## PART IV: MACHINES, MATERIALS AND MANUFACTURING



## PART V: ENERGY AND ENVIRONMENT

# Key Technologies for the 21st Century

# Introduction: The Uncertainties of Technological Innovation

*Even the greatest ideas and inventions can flounder, whereas more modest steps forward sometimes change the world.*

• • •

John Rennie

"The future is not what it used to be," wrote the poet Paul Valéry decades ago, and it would not be hard to share in his disappointment today. As children, many of us were assured that we would one day live in a world of technological marvels. And so we do—but, by and large, not the ones foretold. Films, television, books and World's Fairs promised that the twilight of the 20th century and the dawn of the 21st would be an era of helpful robot servants, flying jet cars, moon colonies, easy space travel, undersea cities, wrist videophones, paper clothes, disease-free lives and, oh, yes, the 20-hour work week. What went wrong?

Few of the promised technologies failed for lack of interest. Nor was it usually the case that they were based on erroneous principles, like the perpetual motion machines that vex patent offices. Quite often, these inventions seemed to work. So why do bad things happen to good technologies? Why do some innovations fall so far short of what is expected of them, whereas others succeed brilliantly?

One recurring reason is that even the most knowledgeable forecasters are sometimes much too optimistic about the short-run prospects for success. Two decades ago, for example, building a self-contained artificial heart seemed like a reasonable, achievable early goal—not a simple chore, of course, but a straightforward one. The heart, after all, is just a four-chambered pump; surely our best biomedical engineers could build a pump! But constructing a pump compatible with the delicate tissues and subtle chemistry of the body has proved elusive. In many ways, surgeons have had far more luck with transplanting organs from one body to another and subduing (through the drug equivalent of brute force) the complex immunologic rejection reactions.

Similarly, from the 1950s through the early 1970s, most artificial-intelligence researchers were smoothly confident of their ability to simulate another organ, the brain. They are more humble these days: although their work has given rise to some narrow successes, such as medical-diagnostic expert systems and electronic chess grandmasters, replicating anything like real human intelligence is now recognized as far more arduous.

The more fundamental problem with most technology predictions, however, is that they are simplistic and, hence, unrealistic. A good technology must by definition be useful. It must be able to survive fierce buffeting by market forces, economic and social conditions, governmental policies, quirky timing, whims of fashion and all the vagaries of human nature and custom. What would-be Nostradamus is prepared to factor in that many contingencies?

Sadly, some inventions are immensely appealing in concept but just not very good in practice. The Buck Rogers–style jetpack is one. With the encouragement of the military, engineers designed and built prototypes during the 1960s. As scene-stealing props in movies such as *Thunderball,* jetpacks embodied tomorrow's soaring high-tech freedom: fly to work, fly to school, fly to the market—

But practical considerations kept jetpacks grounded. The weight of the fuel almost literally sank the idea. The amount required to fly an appreciable distance rapidly became impractical to attach to a user's back. The packs also did not maneuver very well. Finally, the military could not define enough missions that called for launching infantry into the air (where they might be easy targets for snipers) to justify the expense of maintaining the program.

To survive, a commercial technology must not only work well, it must compete in the marketplace. During the 1980s, many analysts thought industrial robotics would take off. Factory managers discovered, however, that roboticizing an assembly line meant more than wheeling the old machines out and the robots in. In many cases, turning to robots would involve completely rethinking (and redesigning) a manufacturing plant's operations. Robots were installed in many factories with good results, particularly in the automobile industry, but managers often found that it was more economical to upgrade with less versatile, less intelligent but more cost-effective conventional machines. (Experts still disagree about whether further advances in robotics will eventually tip this balance.)

Many onlookers thought silicon-based semiconductors would be replaced by faster devices made of new materials, such as gallium arsenide, or with new architectures, such as superconducting Josephson junction switches. The huge R&D base associated with silicon, however, has continued to refine and improve the existing technology. Result: silicon will almost certainly remain the semiconductor of choice for most products for at least as long as the current chip-making technology survives. Its rivals are finding work, too, but in specialized niche applications.

One projected commercial payoff of the space program is supposed to be the development of orbiting manufacturing facilities. In theory, under weightless conditions, it should be possible to fabricate ball bearings, grow semiconductor crystals and purify pharmaceuticals without imperfections caused by gravity. Yet the costs associated with spaceflight remain high, which means that building these factories in space and lofting raw materials to them would be neither easy nor inexpensive. Moreover, improvements in competing ground-based technologies are continuing to eat away at the justification for building the zero-gravity facilities.

Government policies and decisions can also influence the development of new technologies. Yawn-inducing federal decisions about standards for electronic devices and the availability of the broadcast spectrum for commercial use indirectly dictate the rate and results of electronic device development. International disputes about who owns the mineral rights to the seafloor sapped the incentive that many nations and corporations had to invest in undersea mining technologies. Competing industrial standards can also stymie progress—witness the wrangles that froze work on high-definition television.

And sometimes the worth of one technology does not really become clear until other small but crucial inventions and discoveries put them in perspective. Personal computers looked like mere curiosities for hobbyists for many years; not until Dan Bricklin and Mitchell Kapor invented the first spreadsheet programs did personal computers stand out as useful business tools. CD-ROMs did not start to become common accessories of PCs until the huge size of some programs, particularly reference works and interactive games, made the optical disks convenient alternatives to cheaper but less capacious floppies.

In short, the abstract quality of an innovation matters not at all. Build a better mousetrap, and the world may beat a path to your door—if it doesn't build a better mouse instead, or tie up your gadget in environmental-impact and animal-cruelty regulations.

Of course, many technologies do succeed wildly beyond anyone's dreams. Transistors, for instance, were at first seen merely as devices for amplifying radio signals and later as sturdier replacements for vacuum tubes. Ho-hum. Yet their solid-state nature also meant they could be mass-produced and miniaturized in ways that vacuum tubes could not, and their reliability meant that larger devices incorporating greater numbers of components would be feasible. (Building the equivalent of a modern computer with vacuum tube switches instead of transistors would be impossible. Not only would its size make it too slow, the tens of millions of tubes would break down so frequently that the machine would be permanently on the fritz.)

Of those advantages, the microelectronic revolution was born. Similar Horatio Alger stories can be told for lasers, fiber optics, plastics, piezoelectric crystals and other linchpins of the modern world. In fact, it is tempting to think that most great innovations are unforeseen, if not unforeseeable. As computer scientists Whitfield Diffie and John McCarthy reminded panelists in the spring of 1995 at a public discussion on the future hosted by SCIENTIFIC AMERICAN, "A technology-of-the-20th-century symposium held in 1895 might not have mentioned airplanes, radio, antibiotics, nuclear energy, electronics, computers or space exploration."

Given the pitfalls of prognostication, why would SCIENTIFIC AMERICAN dare to prepare a book on key technologies of the 21st century? First, technology and the future have always been the province of the magazine. When SCIENTIFIC AMERICAN was founded in 1845, the industrial revolution was literally still gathering steam. Those were the days before the birth of Edison, before Darwin's *On the Origin of the Species,* before the germ theory of disease, before the invention of cheap steel, before the discovery of x-rays, before Mendel's laws of genetics and Maxwell's equations of electromagnetism. The magazine has had the privilege of reporting on all the major technological advances since that time. The editors of SCIENTIFIC AMERICAN could think of no more fitting way to celebrate its 150th birthday than by taking a peek ahead.

Second, to paraphrase Valéry, the future is now not even *when* it used to be. The early decades of the new century—make that the new millennium—will be when the technologies that now exist and look most promising either flourish or wither on the vine.

In selecting technologies to include in this book, the editors decided to forsake the purely fabulous and concentrate on those that seemed most likely to have strong, steady, enduring effects on day-to-day life. What, some readers may exclaim, no faster-than-light starships? Immortality pills? U-Clone-'Em personal duplication kits? Sorry, but no, not here. In the words of that famous oracle and child's toy the Magic 8 Ball: "Reply hazy, try again."

Naturally, this book makes no pretense of being an exhaustive list of all the technologies that will contribute powerfully to the years ahead. Any attempt to make it one would have sacrificed useful detail for nominal thoroughness. Our more modest intention is only to convey the excitement and real rate of substantive progress in many pivotal fields.

The truth is that as technologies pile on technologies at an uneven pace, it becomes impossible to predict precisely what patterns will emerge. Can anyone today truly foresee what the world will be like if, for example, genetic engineering matures rapidly to its full potential? If organisms can be tailored to serve any function (even becoming living spaceships, as Freeman J. Dyson seems to hint in Chapter 9, "21st-Century Spacecraft"), can anyone guess what a 21st-century factory will look like?

New technologies also pose moral dilemmas, economic challenges, personal and social crises. For example, after the Human Genome Project is completed in a decade or so, the genetic foundations of any biological question will become transparent to investigation. The controversial genetic aspects of intelligence, violence and other complex traits will then be available for direct scrutiny—and, conceivably, manipulation. How much will that transform the basis and practice of medicine, law and government? So in addition to articles on the nuts and bolts of technological development, readers will find here more essayistic commentaries that meditate on the consequences (both good and bad) of the work in progress.

Perceptive readers will also note that some of the authors of this book's chapters implicitly or explicitly disagree with one another; they do not share a consensus on tomorrow. It is precisely out of the tensions between differing predictions that the real future will pull itself together.

# PART I
# INFORMATION TECHNOLOGIES

# Microprocessors in 2020

*Every 18 months microprocessors double in speed. By the year 2020, one computer will be as powerful as all those in Silicon Valley today.*

• • •

David A. Patterson

When I first read the table of contents of this book, I was struck by how many chapters addressed computers in the 21st century in some way. Unlike many other technologies that fed our imaginations and then faded away, the computer has transformed our society. There can be little doubt that it will continue to do so for many decades to come. The engine driving this ongoing revolution is the microprocessor. These silicon chips have led to countless inventions, such as portable computers and fax machines, and have added intelligence to modern automobiles and wristwatches. Astonishingly, their performance has improved 25,000 times over since their invention only 25 years ago.

I have been asked to describe the microprocessor of 2020. Such predictions in my opinion tend to overstate the worth of radical, new computing technologies. Hence, I boldly predict that changes will be evolutionary in nature, and not revolutionary. Even so, if the microprocessor continues to improve at its current rate, I cannot help but suggest that by the year 2020 these chips will empower revolutionary software to compute wonderful things.

## Smaller, Faster, Cheaper

Two inventions sparked the computer revolution. The first was the so-called stored program concept. Every computer system since the late 1940s has adhered to this model, which prescribes a processor for crunching numbers and a memory for storing both data and programs. The advantage in such a system is that, because stored programs can be easily interchanged, the same hardware can perform a variety of tasks. Had computers not been given this flexibility, it is probable that they would not have met with such widespread use. Also, during the late 1940s, researchers invented the transistor. These silicon switches were much smaller than the vacuum tubes used in early circuitry. As such, they enabled workers to create smaller—and faster—electronics.

More than a decade passed before the stored program design and transistors were brought together in the same machine, and it was not until 1971 that the most significant pairing—the Intel 4004—came about. This processor was the first to be built on a single silicon chip, which was no larger than a child's fingernail. Because of its tiny size, it was dubbed a microprocessor. And because it was a single chip, the Intel 4004 was the first processor that could be made inexpensively in bulk.

The method manufacturers have used to mass-produce microprocessors since then is much like baking a pizza: the dough, in this case silicon, starts thin and round. Chemical toppings are added, and the assembly goes into an oven. Heat transforms the toppings into transistors, conductors and insulators. Not surprisingly, the process—which is repeated perhaps 20 times—is considerably more demanding than baking a pizza. One dust particle can damage the tiny transistors (see Figure 1.1). So, too, vibrations from a passing truck can throw the ingredients out of alignment, ruining the end product. But provided that does not happen, the resulting wafer is divided into individual pieces, called chips, and served to customers.

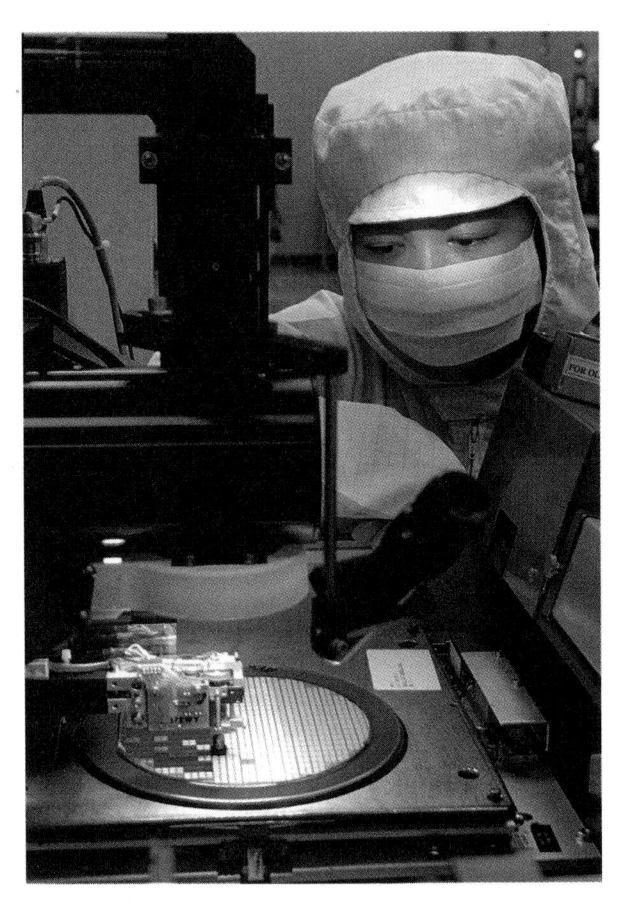

**Figure 1.1 CLEAN ROOMS, where wafers are made, are designed to keep human handling and airborne particles to a minimum. A single speck of dust can damage a tiny transistor.**

Although this basic recipe is still followed, the production line has made ever cheaper, faster chips over time by churning out larger wafers (see Figure 1.2) and smaller transistors. This trend reveals an important principle of microprocessor economics: the more chips made per wafer, the less expensive they are. Larger chips are faster than smaller ones because they can hold more transistors. The recent Intel P6, for example, contains 5.5 million transistors and is much larger than the Intel 4004, which had a mere 2,300 transistors. But larger chips are also more likely to contain flaws. Balancing cost and performance, then, is a significant part of the art of chip design.

Most recently, microprocessors have become more powerful, thanks to a change in the design approach. Following the lead of researchers at universities and laboratories across the U.S., commercial chip designers now take a quantitative approach to computer architecture. Careful experiments precede hardware development, and engineers use sensible metrics to judge their success. Computer companies acted in concert to adopt this design strategy during the 1980s, and as a result, the rate of improvement in microprocessor technology has risen from 35 percent a year only a decade ago to its current high of approximately 55 percent a year, or almost 4 percent each month. Processors are now three times faster than had been predicted in the early 1980s; it is as if our wish was granted, and we now have machines from the year 2000.

## Pipelined, Superscalar and Parallel

In addition to progress made on the production line and in silicon technology, microprocessors have benefited from recent gains on the drawing board. These breakthroughs will undoubtedly lead to further advancements in the near future. One key technique is called pipelining. Anyone who has done laundry has intuitively used this tactic. The nonpipelined approach is as follows: place a load of dirty clothes in the washer. When the washer is done, place the wet load into the dryer. When the dryer is finished, fold the clothes. After the clothes

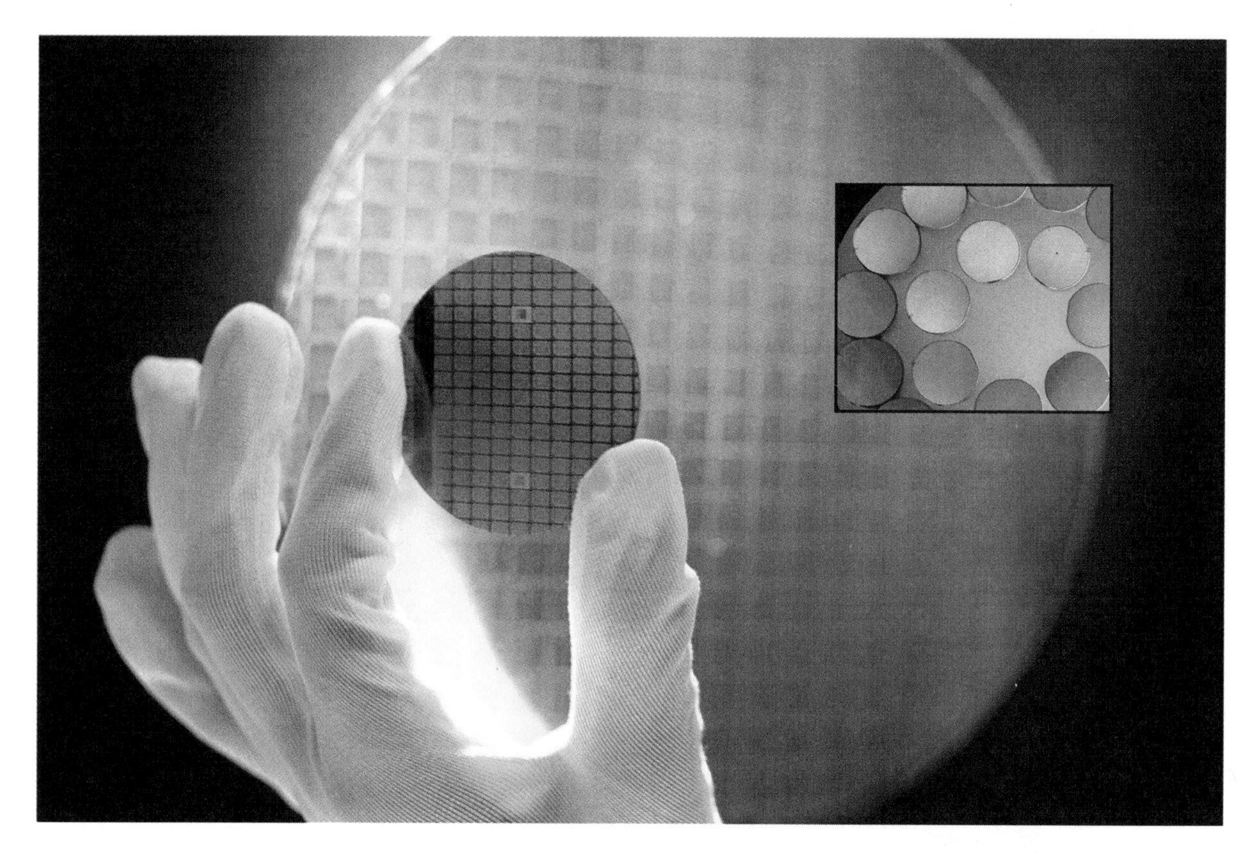

**Figure 1.2 SILICON WAFERS today (*background*) are much larger but hold only about half as many individual chips as did those of the original microprocessor, the Intel 4004 (*foreground*). The dies can be bigger in part because the manufacturing process (*one stage shown in inset*) is cleaner.**

are put away, start all over again. If it takes an hour to do one load this way, 20 loads take 20 hours.

The pipelined approach is much quicker. As soon as the first load is in the dryer, the second dirty load goes into the washer, and so on. All the stages operate concurrently. The pipelining paradox is that it takes the same amount of time to clean a single dirty sock by either method. Yet pipelining is faster in that more loads are finished per hour. In fact, assuming that each stage takes the same amount of time, the time saved by pipelining is proportional to the number of stages involved. In our example, pipelined laundry has four stages, so it would be nearly four times faster than nonpipelined laundry. Twenty loads would take roughly five hours.

Similarly, pipelining makes for much faster microprocessors. Chip designers pipeline the instructions, or low-level commands, given to the hardware. The first pipelined microprocessors used a five-stage pipeline. (The number of stages completed each second is given by the so-called clock rate. A personal computer with a 100-megahertz clock then executes 100 million stages per second.) Because the speedup from pipelining equals the number of stages, recent microprocessors have adopted eight or more stage pipelines. One 1995 microprocessor uses this deeper pipeline to achieve a 300-megahertz clock rate. As machines head toward the next century, we can expect pipelines having even more stages and higher clock rates.

Also in the interest of making faster chips, designers have begun to include more hardware to process more tasks at each stage of a pipeline. The buzzword "superscalar" is commonly used to describe this approach. A superscalar laundromat, for example, would use a professional machine that could, say, wash three loads at once. Modern superscalar microprocessors try to perform anywhere from three to six instructions in each stage. Hence, a 250-megahertz, four-way superscalar microprocessor can execute a billion instructions per second. A 21st-century microprocessor may well launch up to dozens of instructions in each stage.

Despite such potential, improvements in processing chips are ineffectual unless they are matched by similar gains in memory chips. Since random-access memory (RAM) on a chip became widely available in the mid-1970s, its capacity has grown fourfold every three years. But memory

## The Limits of Lithography

Although I predict that microprocessors will continue to improve rapidly, such a steady advance is far from certain. It is unclear how manufacturers will make tinier, faster transistors in the years to come. The photolithographic methods they now use are reaching serious limits. If the problem is not resolved, the progress we have enjoyed for decades will screech to a halt.

In photolithography, light is used to transfer circuit patterns from a quartz template, or mask, onto the surface of a silicon chip. The technique now fashions chip features that are some 0.35 micron wide. Making features half as wide would yield transistors four times smaller, since the device is essentially two-dimensional. But it seems impossible to make such tiny parts using light; the light waves are just too wide. Many companies have invested in finding ways to substitute smaller x-rays for light waves. To date, however, x-rays have not succeeded as a way to mass-produce state-of-the-art chips.

**PHOTOMASKS are reduced and projected onto silicon wafers to make circuits.**

Other proposals abound. One hope is to deploy the electron beams used to create quartz masks to pattern silicon wafers. The thin stream of charged particles could trace each line in a circuit diagram, one by one, directly onto a chip. The catch is that although this solution is feasible, it is unreasonably slow for commercial use and would therefore prove costly. Compared with photolithography, drawing with an electron beam is analogous to rewriting a letter by hand instead of photocopying it.

Technical hurdles aside, any improvements in microprocessors are further threatened by the rising cost of semiconductor manufacturing plants. At $1 billion to $2 billion, these complexes now cost 1,000 times more than they did 30 years ago. Buyers and sellers of semiconductor equipment follow the rule that halving the minimum feature size doubles the price. Clearly, even if innovative methods are found, the income generated by the sale of smaller chips must double to secure continued investments in new lines. This pattern will happen only by making more chips or by charging more for them.

Today there are as many companies that have semiconductor lines as there are car companies. But increasingly few of them can afford the multibillion-dollar cost of replacing the equipment. If semiconductor equipment manufacturers do not offer machinery that trades off, say, the speed of making a wafer for the cost of the equipment, the number of companies making state-of-the-art chips may shrink to a mere handful. Without the spur of competition, once again, the rapid pace of improvement may well slow down.

## And After 2020?

With decades of innovative potential ahead of them, conventional microelectronic designs will dominate much of the 21st century. That trend does not discourage many laboratories from exploring a variety of novel technologies that might be useful in designing new generations of computers and microelectronic devices. In some cases, these approaches would allow chip designs to reach a level of miniaturization unattainable through anything like conventional lithography techniques. Among the ideas being investigated are:

- **Quantum dots and other single-electron devices.** Quantum dots are molecular arrays that allow researchers to trap individual electrons and monitor their movements. These devices can in theory be used as binary registers in which the presence or absence of a single electron is used to represent the 0 or 1 of a data bit. In a variation on this scheme, laser light shining on atoms could switch them between their electronic ground state and an excited state, in effect flipping the bit value.

One complication of making the transistors and wires extremely small is that quantum-mechanical effects begin to disrupt their function. The logic components hold their 0 or 1 values less reliably because the locations of single electrons become hard to specify. Yet this property could be exploited: Seth Lloyd of the Massachusetts Institute of Technology and other researchers are studying the possibility of developing quantum computing techniques, which would capitalize on the nonclassical behavior of the devices.

- **Molecular computing.** Instead of making components out of silicon, some investigators are trying to develop data storage systems using biological molecules. Robert L. Birge of Syracuse University, for example, is examining the computational potential of molecules related to bacteriorhodopsin, a pigment that alters its configuration in response to light. One advantage of such a molecule is that it could be used in an optical computer, in which streams of photons would take the place of electrons. Another is that many of these molecules might be synthesized by microorganisms, rather than fabricated in a factory. According to some estimates, photonically activated biomolecules could be linked into a three-dimensional memory system that would have a capacity 300 times greater than today's CD-ROMs.

**QUANTUM DOT (*purple*) in this semiconductor structure traps electrons.**

- **Nanomechanical logic gates.** In these systems, tiny beams or filaments only one atom wide might be physically moved, like Tinkertoys, to carry out logical operations (see Chapter 13, "Self-Assembling Materials," by George M. Whitesides).
- **Reversible logic gates.** As the component density on chips rises, dissipating the heat generated by computations becomes more difficult. Researchers at Xerox PARC, the IBM Thomas J. Watson Research Center and elsewhere are therefore checking into the possibility of returning capacitors to their original state at the end of a calculation. Because reversible logic gates would in effect recapture some of the energy expended, they would generate less waste heat.

—*John Rennie, Scientific American*

speed has not increased at anywhere near this rate. The gap between the top speed of processors and the top speed of memories is widening.

One popular aid is to place a small memory, called a cache, right on the microprocessor itself. The cache holds those segments of a program that are most frequently used and thereby allows the processor to avoid calling on external memory chips much of the time. Some newer chips actually dedicate as many transistors to the cache as they do to the processor itself. Future microprocessors will allot even more resources to the cache to better bridge the speed gap.

The Holy Grail of computer design is an approach called parallel processing, which delivers all the benefits of a single fast processor by engaging many inexpensive ones at the same time. In our analogy, we would go to a laundromat and use 20 washers and 20 dryers to do 20 loads simultaneously. Clearly, parallel processing is an expensive solution for a small workload. And writing a program that can use 20 processors at once is much harder than distributing laundry to 20 washers. Indeed, the program must specify which instructions can be launched by which processor at what time.

Superscalar processing bears similarities to parallel processing, and it is more popular because the hardware automatically finds instructions that launch at the same time. But its potential processing power is not as large. If it were not so difficult to write the necessary programs, parallel processors could be made as powerful as one could afford. For the past 25 years, computer scientists have predicted that the programming problems will be overcome. In fact, parallel processing is practical for only a few classes of programs today.

In reviewing old issues of SCIENTIFIC AMERICAN, I have seen fantastic predictions of what computers would be like in 1995. Many stated that optics would replace electronics; computers would be built entirely from biological materials; the stored program concept would be discarded. These descriptions demonstrate that it is impossible to foresee what inventions will prove commercially viable and go on to revolutionize the computer industry. In my career, only three new technologies have prevailed: microprocessors, random-access memory and optical fibers. And their impact has yet to wane, decades after their debut.

Surely one or two more inventions will revise computing by the year 2020. My guess, though, is that the stored program concept is too elegant to be easily replaced. I believe future computers will be much like machines of the past, even if they are made of very different stuff. I do not think the microprocessor of 2020 will be startling to people from our time, although the fastest chips may be much larger than the very first wafer, and the cheapest chips may be much smaller than the original Intel 4004.

## IRAMs and Picoprocessors

Pipelining, superscalar organization and caches will continue to play major roles in the advancement of microprocessor technology, and if hopes are realized, parallel processing will join them. What will be startling is that microprocessors will probably exist in everything from light switches to pieces of paper. And the range of applications these extraordinary devices will support, from voice recognition to virtual reality, will very likely be astounding.

Today microprocessors and memories are made on distinct manufacturing lines, but it need not be so. Perhaps in the near future, processors and memory will be merged onto a single chip, just as the microprocessor first merged the separate components of a processor onto a single chip. To narrow the processor-memory performance gap, to take advantage of parallel processing, to amortize the costs of the line and simply to make full use of the phenomenal number of transistors that can be placed on a single chip, I predict that the high-end microprocessor of 2020 will be an entire computer.

Let's call it an IRAM, standing for intelligent random-access memory, since most of the transistors on this merged chip will be devoted to memory. Whereas current microprocessors rely on hundreds of wires to connect to external memory chips, IRAMs will need no more than computer network connections and a power plug. All input-output devices will be linked to them via networks. If they need more memory, they will get more processing power as well, and vice versa—an arrangement that will keep the memory capacity and processor speed in balance. IRAMs are also the ideal building block for parallel processing. And because they would require so few external connections, these chips could be extraordinarily small. We may well see cheap "picoprocessors" that are smaller than the ancient Intel 4004. If parallel processing succeeds, this sea of transistors could also be used by multiple processors on a single chip, giving us a micromultiprocessor.

Today's microprocessors are almost 100,000 times faster than their Neanderthal ancestors of the 1950s, and when inflation is considered, they cost 1,000 times less. These extraordinary facts explain why computing plays such a large role in our world now. Looking ahead, microprocessor performance will easily keep doubling every 18 months through the turn of the century. After that, it is hard to bet against a curve that has outstripped all expectations. But it is plausible that we will see improvements in the next 25 years at least as large as those seen in the past 50. This estimate means that one desktop computer in 2020 will be as powerful as all the computers in Silicon Valley today. Polishing my crystal ball to look yet another 25 years ahead, I see another quantum jump in computing power. The implications of such a breathtaking advance are limited only by our imaginations.

# Wireless Networks

*In the decade ahead, they will deliver personalized communications to people on the go and basic service to many who still lack telephones.*

• • •

George I. Zysman

Near the end of the 19th century a young man named Guglielmo Marconi connected a spark emitter to a short antenna and sent a burst of radio waves through the air to a simple receiver. It responded by ringing a bell, signaling the birth of a technology that promised to allow people to communicate across distances while in motion. In the closing decades of the 20th century, several waves of innovation have made wireless communications the fastest-growing segment of the global telecommunications industry (see Figure 2.1).

Wireless networks are proliferating rapidly, going digital and harnessing "intelligent network" technology to locate and identify roaming subscribers and to customize the services they receive. An intelligent network consists of a distributed signaling network of switches, databases and dedicated computer servers that is separate from, yet intimately connected to, the transport networks on which subscribers' voice calls and data actually flow. This architectural framework, which has been refined over the past 30 years to support such services as 800-number calling, caller identification and "911," will soon make personalized communications services as portable as a pocket telephone.

As advances in microelectronics, digital radio, signal processing and network software converge in the marketplace, portable telephones are getting smaller, smarter and less expensive. Some are taking on new forms, such as the wireless handheld computers called personal digital assistants (PDAs), so that they can handle text and graphics as well as audio messages; video is not far behind. Increasingly, the software running on "smart" terminals—typified by the graphical user interfaces and intelligent software agents available today in PDAs—will work hand in hand with intelligent networks to enhance portable communications.

Since 1990 the demand for wireless services has risen beyond all expectations. In 1983 some industry analysts predicted that fewer than one million Americans would use cellular services by the year 2000. Currently more than 20 million do. Cellular services now spearhead the market penetration of wireless communications, as the number of cellular users grows annually by approximately 50 percent

**Figure 2.1 WIRELESS NETWORKS based on cellular technology will be the first infrastructure to provide telephone service in some places, such as on this ranch in Argentina.**

in North America, 60 percent in western Europe, 70 percent in Australia and Asia, and more than 200 percent in South America's largest markets.

Analysts now project that by 2001, three quarters of the households in the U.S. and nearly half a billion people worldwide will subscribe to a wireless service of some kind. The Federal Communications Commission has raised nearly $8 billion in the past year, with more to come, by auctioning licenses to use emerging technologies and radio spectrum around the frequency of two gigahertz to provide a new set of wireless capabilities known as personal communications services, or PCS. The terms of these auctions require licensees to move quickly to install the infrastructure needed to provide PCS. The magnitude of the investment that has been made by PCS licensees and equipment manufacturers is a measure of the industry's confidence in the projected market demand. This phenomenon is not limited to the U.S. Before the century is out, service providers in Europe, Japan, Thailand, Singapore, Malaysia, China, Australia, New Zealand and India all reportedly plan to have PCS systems up and running.

The growth of the wireless market has increased pressure on regulatory bodies to allocate more spectrum and on service providers to use spectrum more efficiently by converting to digital technology.

## The Switch to Digital

The present analog standards used by most cellular systems encode voices and even digital data into continuous variations of a carrier wave, which are then decoded by the receiver. Already many cellular service providers are converting their networks to one of several digital standards that translate voices and data into a bit stream, which is sent in waveforms that represent discrete pulses. Compared with their analog counterparts, digital systems can both expand the capacity of the medium and compress the messages it carries.

Most cellular and PCS networks will soon use one of the digital air interface standards—different ways of sharing the limited spectrum among many users at a time—that are now vying for acceptance. Whether one will eventually win out remains to be seen. In the most likely scenario, intelligent base stations and dual-mode terminals will adapt to a patchwork of multiple-access air interface standards spread across the wireless landscape. But all the leading digital air interface standards offer a similar benefit: the ability to pack more bits of conversations into a slice of spectrum than an analog system can.

Once wireless service providers switch to digital, they can further increase the number of customers served by employing compression techniques, which are improving steadily. A stream of eight kilobits per second can transmit good-quality speech; better quality, delivered not long ago at the rate of 32 kilobits per second, now requires only 13.

Service providers can also keep ahead of demand by shrinking the size of each cell—the area covered by a single base station—in crowded areas. It is much easier to add small cells with digital standards, since they provide error correction and help the receivers resolve interference between adjacent cells.

The move to all-digital technology is driving communications terminals toward greater functionality, smaller size and lower power. Portable telephones and other wireless devices are essentially miniature computers with some extra electronics to transmit and receive radio signals. As such, they are susceptible to Moore's Law, an axiom first postulated by Gordon Moore, cofounder of Intel, in 1965. It observes that the performance of mass-produced microchips doubles every 18 months or so.

Every year and a half the digital chips needed to run a wireless terminal or base station shrink by about 50 percent. Already cellular telephones are slipped into pockets. Soon they could be strapped onto wrists. Analog base stations that currently require towers, real estate and air-conditioned shacks will eventually be replaced by inconspicuous digital base stations serving minicells. Microcell systems deployed to cover very small areas may even become the size of a smoke detector.

Over the next few years, cable television operators will begin adding base stations to their fiber-optic and coaxial-cable networks, carrying telephone traffic on unused cable channels and supplying wireless access to neighborhoods in competition with other local access providers. If they use the same air interface standard as a local cellular carrier, their telephone customers could be able to place calls over the cellular network, and vice versa. Power companies, which own ubiquitous grids of communications as well as power facilities, are entertaining similar thoughts.

## Data on the Air

Although portable phones and pagers are certainly convenient—after all, two out of three business calls still end in "telephone tag"—new devices and network systems that can transmit and receive text and images over the air will have a larger impact, in the long term, on the way people communicate. Built-in radio modems can link laptops, PDAs and other handheld digital devices over today's predominantly analog cellular networks, and there are several dedicated wireless data networks in service. Digital cellular networks for mobile and packet data services are beginning to offer other alternatives. Licenses to provide "narrowband PCS"—two-way paging and moderate-speed data messaging services at frequencies around 900 megahertz—were awarded by the FCC through auctions held in 1994.

First-generation handheld wireless computers did not catch on, perhaps because they were somewhat awkward and had too little functionality for the price. But as people of every age and income grow increasingly familiar with electronic mail, commercial on-line services and the Internet, it stands to reason that they will want access to the information these media offer at any time, according to need or whim—not just when they happen to be sitting at a computer.

It is possible to adapt old-fashioned devices to receive newfangled messages. Several companies offer to filter and forward E-mail, one line at a time, to your alphanumeric pager via satellite. Artificial-speech chips are now sophisticated and inexpensive enough that they could be put into telephones, so that they would read incoming text aloud. But more versatile devices will require displays and reasonably fast transmission rates to handle graphics and text.

Analog network technology limits data communications via modem to relatively slow speeds: 14.4 kilobits per second or less. Digital networking will help somewhat by eliminating conversions between analog and digital formats; it will not necessarily increase data rates for standard service beyond the equivalent of a telephone line.

Eventually, even the boost from the present digital standards and compression techniques will prove insufficient to provide enough room to pass data-intensive messages among hundreds of millions of customers. By the turn of the century, wireless faxes will be common, and video mail could be widely used throughout wireless as well as wireline networks. If spectrum is allocated to accommodate the resulting deluge of images, it will probably be at frequencies of around 30 to 40 gigahertz, although some video services may be available at 2.5 gigahertz. Because radio signals at these short wavelengths behave like light, buildings and even foliage block them, and coverage is essentially limited to line of sight. So wireless broadband services would be fixed rather than mobile, and service providers would need to shrink cells further to serve large numbers of subscribers.

Air interface standards will also probably evolve to accommodate broadband data and video transmission. The first generation of digital standards presumes a circuitlike connection between two devices for an entire call. That arrangement is better suited to telephone calls than to surfing the Net. The Internet uses standards based on routing individually addressed packets of data.

Telephone companies designing fiber-optic networks to carry interactive video services plan to connect them with broadband packet switches—using a technology called asynchronous transfer mode, or ATM—that can shunt packets of data, voice and video along the appropriate paths at extremely fast rates. Wireless networks will follow this trend when it becomes the most efficient way to combine voice and multimedia services. It is possible, in fact, that wireless service providers and cable TV companies may lead the shift to ATM. The timing will depend on existing network assets, technological plans and investment strategies.

## Intelligence in Motion

As portable telephones grow in popularity, they will also tend to offer more features. Ten years from now you will almost certainly still be able to buy a simple wireless device that is limited to voice calls; it will probably be quite inexpensive. More common, however, will be devices that can handle faxes and video and run software applications. Doubtless there will be many styles with widely varying capabilities.

This could cause problems. The smarter a device must be, the greater the risk that its complexity will baffle the user. And different models will work differently. Picking up a friend's telephone to make a call might leave you staring at a completely unfamiliar user interface.

One way to simplify the operation of handheld wireless gadgets and make them more user-friendly is to place some of the intelligence needed to accomplish useful tasks in the network. Intelligent network services available now can, for example, forward calls automatically to a subscriber's car, office, home, portable telephone or voice mail. Software architectures and systems that facilitate collaboration between intelligent devices and intelligent networks will improve the accuracy of personal mobility services—ringing the right device on the first try—and will make much more sophisticated interactions possible.

At AT&T, the vision of PCS is that it will deliver the right service to the right location and device without any intervention by the caller or the subscriber. The trick is keeping track of where subscribers are. One solution is to use personal numbers: one per person, instead of the three, five, seven or more that today address different devices and network locations where a person might be reached. Register with the network, and the network will translate your personal number, as dialed by a caller, to the number for the appropriate local device or mailbox, depending on the type of call or message and on your service preferences.

Smart cards offer another solution. If all telephones were equipped with a card reader, you could insert your card in the nearest telephone, even if it belonged to someone else, and register your presence with the network. It would then direct all your calls—or perhaps only those from a special list of priority numbers—to that telephone. Or you could simply send the network a normal daily schedule and use the card to register exceptions. People who would rather not be found would register that with the network, too.

Smart cards might also help get around the complexity of using sophisticated telephones. Currently when one travels to another country, dialing 911 does not get an emergency operator; 011 is rarely the correct prefix for international calls. Such country-specific codes may proliferate as devices become more complex. But an interface description could be stored on a smart card. Insert that card in a strange device, and it could reconfigure itself to work just like your own. You would not have to learn how to use a plethora of new devices; instead they would learn how to work with you.

Smart cards have been technologically demonstrated. Whether the business and personal privacy issues they raise can be settled is another matter. Companies in a wide variety of businesses see uses for smart cards, but each has an interest in controlling the personal information that will be stored on them.

There are other, less controversial ways to track callers and to provide location-specific services. Cellular systems can already locate a caller's position to within a few square miles. A more precise alternative would be to equip devices with Global Positioning System receivers, which can often pinpoint their location to within about 100 feet using

## Wireless Telephony for Developing Countries

The same technologies that provide flip-phone service to people on the go can provide basic dial tone to regions where it has never been heard before. This is much of the planet: about half the people alive today have never made a telephone call. A number of developing countries have expressed a desire to leapfrog over a generation or two of network technology, using wireless infrastructure to jump-start telephone service.

These countries see two advantages in wireless telephone systems. The first is cost. Many of the facilities necessary for a conventional wire-line access network are typically engineered for 20 to 30 years of service expansion. As a result, the cost of building such a network where no infrastructure exists may be prohibitively high.

**PARTS OF ARGENTINA inaccessible to wire-line telephone service are now served by a wireless network built in less than six months.**

With a "fixed wireless" network geared for access rather than mobility, a service provider can cover a large region with base stations and a single unit for switching and control—a significantly smaller investment. Subscribers can then connect to the global network using portable telephones, wireless public telephones, or terminals mounted on buildings and wired to conventional telephones. As the number of subscribers grows, the service provider can easily add more base stations to split the area covered by the network into smaller segments.

The second benefit is time. Wireless networks can be installed in months rather than the years required to install copper wires. Argentina, for example, announced in February 1994 that it was awarding licenses for the entire country to CTI, a GTE-led consortium. By May an 800-cell fixed wireless network that AT&T built for CTI was up and running. It can serve as many as 160,000 subscribers.

Telecommunications, many have observed, constitutes the main infrastructure for the global economy. Where there is a potential for revenue, wireless technology could offer a bootstrap to participation in that market for many regions of the world that might otherwise be excluded.

signals from a constellation of orbiting satellites. Equipping cell sites to locate devices by triangulation could be less expensive than using the GPS; this third method might even be more accurate.

In emergencies, such techniques could help 911 services reach callers who do not know where they are. More routinely, networks could offer customers intelligent access to interactive data services in a kind of "information mall" that pulls together services from many independent companies. Arriving in an unfamiliar town, a traveler might use her PDA to request a list of nearby Italian restaurants from the network. The network could pass this request to a program run by the Italian Restaurant Association of America. The listing it generates might then be sent to an interface program provided by a third company before appearing on the traveler's screen. Her PDA would not have to do much work at all.

Such advanced services will have to wait for the business to develop. One of the biggest hurdles is billing: every company wants to be the one that bonds with customers and collects information about them, since that can lead to new business opportunities.

Ultimately, the distinction between wireless and wire-line networks will recede to the vanishing point. Portable devices will be no more difficult to use than their wired counterparts and will offer equivalent performance and services. Hiding the complexity of wireless networking technology from the people it serves is admittedly a challenge. But the technology is here and will ultimately be put to its proper use—making itself invisible.

# Artificial Intelligence

*A crucial storehouse of commonsense knowledge is now taking shape.*

• • •

Douglas B. Lenat

One of the most frustrating lessons computers have taught us time and time again is that many of the actions we think of as difficult are easy to automate—and vice versa. In 1944 dozens of people spent months performing the calculations required for the Manhattan Project. Today the technology to do the same thing costs pennies. In contrast, when researchers met at Dartmouth College in the summer of 1956 to lay the groundwork for artificial intelligence (AI), none of them imagined that 40 years later we would have come such a short distance toward that goal.

Indeed, what few successes artificial intelligence has had point up the weakness of computerized reasoning as much as they do its narrow strengths. In 1965, for example, Stanford University's Dendral project automated sophisticated reasoning about chemical structures; it generated a list of all the possible three-dimensional structures for a compound and then applied a small set of simple rules to select the most plausible ones. Similarly, in 1975 a program called Mycin surpassed the average physician in the accuracy with which it diagnosed meningitis in patients. It rigorously applied the criteria that expert clinicians had developed over the years to distinguish among the three different causes of the disease. Such tasks are much better suited to a computer than to a human brain because they can be codified as a relatively small set of rules to follow; computers can run through the same operations endlessly without tiring.

Meanwhile many of the tasks that are easy for people to do—figure out a slurred word in a conversation or recognize a friend's face—are all but impossible to automate, because we have no real idea of how we do such things. Who can write down the rules for recognizing a face?

As a result, amid the explosive progress in computer networking, user interface agents and hardware, artificial intelligence appears to be an underachiever. After initial gains led to high expectations during the late 1970s and early 1980s, there was a bitter backlash against AI in both industry and government. Ironically, in September 1984, just as the mania was at its peak, I wrote an article for SCIENTIFIC AMERICAN ("Computer Software for Intelligent Systems") in which I dared to be fairly pessimistic about the coming decade. And now that the world has all but given up on the AI dream, I believe that artificial intelligence stands on the brink of success.

My dire predictions arose because the programs that fueled AI hype were not savants but idiot savants. These so-called expert systems were often right, in the specific areas for which they had been built, but they were extremely brittle. Given even a simple problem just slightly beyond their expertise, they would usually get a wrong answer, without any recognition that they were outside their range of competence. Ask a medical program about a rusty old car, and it might blithely diagnose measles.

Furthermore, these programs could not share their knowledge. Mycin could not talk to programs that diagnosed lung diseases or advised doctors on cancer chemotherapy, and none of the medical programs could communicate with expert scheduling systems that might try to allocate hospital resources. Each represented its bit of the world in idiosyncratic and incompatible ways because developers had cut corners by incorporating many task-specific assumptions. This is still the case today.

## No Program Is an Island

People share knowledge so easily that we seldom even think about it. Unfortunately, that makes it all the more difficult to build programs that do the same. Many of the prerequisite skills and assumptions have become implicit through millennia of cultural and biological evolution and through universal early childhood experiences. Before machines can share knowledge as flexibly as people do, those prerequisites need to be recapitulated somehow in explicit, computable forms.

For the past decade, researchers at the CYC project in Austin, Tex., have been hard at work doing exactly that. Originally, the group examined snippets of news articles, novels, advertisements and the like and for each sentence asked: "What did the writer assume that the reader already knew?" It is that prerequisite knowledge, not the content of the text, that had to be codified. This process has led the group to represent 100,000 discrete concepts and about one million pieces of commonsense knowledge about them.

Many of these entities—for example, "Body Of Water"—do not correspond to a single English word. Conversely, an innocuous word such as "in" turns out to have two dozen meanings, each corresponding to a distinct concept. The way in which you, the reader, are in a room is different from the way the air is in the room, the way the carpet is in the room, the way the paint on the walls is in the room and the way a letter in a desk drawer is in the room. Each way that something can be "in" a place has different implications—the letter can be removed from the room, for instance, whereas the air cannot. Neither the air nor the letter, however, is visible at first glance to someone entering the room.

## What Everyone Knows

Most of these pieces of knowledge turned out not to be facts from an almanac or definitions from a dictionary but rather common observations and widely held beliefs. CYC had to be taught how

**Figure 3.1 COMMON SENSE undergirds even the simplest tasks that people perform. A person searching for pictures of wet people, for instance, can scan through a vast list of captions and pick out likely candidates by bringing commonsense knowledge of the world to bear, even if the image description contains no synonym of "wet." A program with access to the same kind of knowledge can also perform this task, running through a short chain of inferences to decide whether a caption is relevant. Yet without those few pieces of everyday knowledge, no amount of reasoning would suffice.**

***Show me a person who is wet.***

***Salvador Garcia finishing a marathon in 1990.***

*"Running a marathon entails running for at least a couple of hours."*

```
(Implies (∈ e runningAMarathon)
        (And (∈ e Running)
        (duration e (IntervalMin (HoursDuration 2)))
        (Implies (performedBy e x) ∈ x Person))))
```

***Salvador Garcia is a person who has been running for more than two hours.***

"Running a marathon entails running for

*"Running entails high exertion."*

```
(Implies (And (∈ e Running) (performedBy e x))
        (levelOfPhysicalExertion x e High))
```

***Salvador Garcia has been doing high exertion for more than two hours.***

"Running a marathon entails running for
"Running entails high exertion."

*"People sweat when doing something at a high exertion level; they generally start sweating within 10 minutes and continue for at least a couple of minutes after they stop that activity."*

```
(Implies (And (∈ x Person)
        (levelOfPhysicalExertion x e High)
        (duration e (IntervalMin (MinutesDuration 10))))
    (ThereExists s
      (And (∈ s Sweating)
          (doneBy s x)
          (temporallySubsumes
           s
           (TimeIntervalFrom
             (DateAfter (StartOf e) (minutesDuration 10))
             (DateAfter (EndOf e) (minutesDuration 2)))))))
```

***Salvador Garcia is sweating.***

"Running a marathon entails running for
"Running entails high exertion."
"People sweat when doing something at a high

*"You're wet when you sweat."*

```
(Implies (And (∈ e Sweating) (doneBy e x))
        (holdsIn e (wetnessOfObject x Wet)))
```

***Salvador Garcia is wet.***

people eat soup, that children are sometimes frightened by animals and that police in most countries are armed.

To make matters even more complicated, many of the observations we incorporated into CYC's knowledge base contradict one another. By the time a knowledge-based program grows to contain more than 10,000 rules—1 percent of CYC's size—it becomes difficult to add new knowledge without interfering with something already present. We overcame this hurdle by partitioning the knowledge base into hundreds of separate microtheories, or contexts. Like the individual plates in a suit of armor, each context is fairly rigid and consistent, but articulations between them permit apparent contradictions among contexts. CYC knows that Dracula was a vampire, but at the same time it knows that vampires do not exist.

Fictional contexts (such as the one for Bram Stoker's novel) are important because they allow CYC to understand metaphors and use analogies to solve problems. Multiple contexts are also useful for reasoning at different levels of detail, for capturing the beliefs of different age groups, nationalities or historic epochs, and for describing different programs, each of which makes its own assumptions about the situation in which it will be used. We can even use all the brittle idiot savants from past generations of AI by wrapping each one in a context that describes when and how to use it appropriately.

The breadth of CYC's knowledge is evident even in a simple data retrieval application we built in 1994, which fetches images whose descriptions match the criteria a user selects. In response to a request for pictures containing seated people, CYC was able to locate this caption: "There are some cars. They are on a street. There are some trees on the side of the street. They are shedding their leaves. Some of them are yellow taxicabs. The New York City skyline is in the background. It is sunny." The program then used its formalized common sense about cars—they have a driver's seat, and cars in motion are generally being driven—to infer that there was a good chance the image was relevant. Similarly, CYC could parse the request "Show me happy people" and deliver a picture whose caption reads, "A man watching his daughter learn to walk." None of the words are synonymous or even closely related, but a little common sense makes it easy to find the connection (see Figure 3.1).

## Ready for Takeoff

CYC is far from complete, but it is approaching the level at which it can serve as the seed from which a base of shared knowledge can grow. Programs that understand natural languages will employ the existing knowledge base to comprehend a wide variety of texts laden with ambiguity and metaphor; information drawn from CYC's readings will augment its concepts and thus enable further extensions. CYC will also learn by discovery, forming plausible hypotheses about the world and testing them. One of the provocative analogies it noticed and explored a few years ago was that between a country and a family. Like people, CYC will learn at the fringes of what it already knows, and so its capacity for education will depend strongly on its existing knowledge.

During the coming decade, researchers will flesh out CYC's base of shared knowledge by both manual and automated means. They will also begin to build applications, embedding common sense in familiar sorts of software appliances, such as spreadsheets, databases, document preparation systems and on-line search agents.

Word processors will check content, not just spelling and grammar; if you promise your readers to discuss an issue later in a document but fail to do so, a warning may appear on your screen. Spreadsheets will highlight entries that are technically permissible but violate common sense. Document retrieval programs will understand enough of the content of what they are searching—and of your queries—to find the texts you are looking for regardless of whether they contain the words you specify.

These kinds of programs will act in concert with existing trends in computer hardware and networks to make computer-based services ever less expensive and more ubiquitous, to build steadily better user models and agent software and to immerse the user deeper in virtual environments. The goal of a general artificial intelligence is in sight, and the 21st-century world will be radically changed as a result. The late Allen Newell, one of the field's founders, likened the coming era to the land of Faery: inanimate objects such as appliances conversing with you, not to mention conversing and coordinating with one another. Unlike the creatures of most fairy stories, though, they will generally be plotting to do people good, not ill.

# Commentary: Virtual Reality

*VR will transform computers into extensions of our whole bodies.*

• • •

Brenda Laurel

Since the hype began in the mid-1980s, virtual reality (VR) has captivated public interest with pictures of people wearing enormous goggles and sensor-laden gloves. The technologies used to immerse people in a computer-generated world will, however, change radically during the coming decade, making the begoggled cybernaut as quaint an image as the undersea explorer in a heavy metal diving helmet.

The important thing about VR is what it does rather than how its effects are achieved: it permits people to behave as if they were somewhere they are not. That place may be a computational fiction or a re-created environment from another place or time. VR transports perceptions by appealing to several senses at once—sight, hearing and touch—and by presenting images that respond immediately to one's movements. The techniques for creating this illusion differ depending on the kind of place being visited and what a user wants to do while there. A pilot in a flight simulator, for example, might need hydraulic actuators to simulate banks and turns, whereas a molecular biologist exploring the bonds between molecules might need particularly fine position sensors and mechanisms to simulate the "feel" of interatomic forces.

Certainly the coming decades will bring dramatic improvements in existing applications through faster, "smarter" computing and improved interface technology. Bulky, head-mounted stereoscopic displays will be replaced by light-weight "glasses" that can superimpose synthesized images on the real world. The encrustations of tracking and sensing devices cybernauts now wear will be integrated into clothing or replaced by video cameras and other sensors that monitor movements and gestures from a distance. Similarly, technologies that simulate the sensations of force, resistance, texture and smell will become available.

Initially, such new equipment will make existing applications work faster and more comfortably. Already people are performing complex, delicate tasks in hazardous environments, such as space or the inside of nuclear reactors. Pilots and astronauts train in VR cockpits that merge three-dimensional graphics with the view out the window and that contain sound systems offering cues about their surroundings. Architects and planners walk through the environments they design to see how it might feel to live and work inside them. (For a speculative foretaste of the wider possibilities, one can go to arcades and amusement parks where people fly combat missions, fight dinosaurs or travel through the human body.)

As software evolves, as computing power increases, VR will be used to present models of all kinds of complex dynamic systems, from personal investments to global economics and from microorganisms to galaxies. During the past decade, the success of scientific visualization has shown how to harness people's ability to see patterns in properly presented data; soon it will be possible to bring multiple senses to bear simultaneously, engendering a response from the mind and the body that will be more than the sum of its parts.

The social uses of VR will also be an important force in its evolution. Even in the simple text-based on-line environments known as MUDs (Multi-User Dimensions), researchers have shown that a sense of place is crucial to communication and community. In the real world, people devote a great deal of energy to creating particular places as a context for social interaction—consider display windows, architecture and interior design. VR will make it possible to carry many of those skills over to cyberspace. As virtual spaces begin taking on a richer, more complex texture, VR will be the foundation of a major transformation in the ethos of computing. People have until now thought of computers as the last stop on the road of mind-body dualism: as close to disembodied thought as the material world permits. Computers generally have no sense organs, nor do they address human senses particularly well. They have evolved as a race of severed heads, doomed by the arcana of their communications mechanisms to make extremely small talk with people who are almost as strange as they are.

VR, in contrast, makes little or no distinction between body and mind. Instead it employs in a new context the bodily senses that evolution has so magnificently prepared. VR is concerned with the nature of the body—how our senses work, how we move around, how we get the feeling of being somewhere and how the sense of presence affects us. It is also concerned with representing the nature of things, both virtual and actual, in ways that reveal their structure, dynamics and potential uses.

Artists, who have always had to think about the interplay between intellectual and physical responses to their work, may play a more pivotal role in the development of VR than technologists, who may be content with the computers as a medium that exclusively addresses the disembodied human intellect. As artists explore the expressive potential of VR, they will grow more adept at representing the subtleties and complexities of experience—from "synthesthesia" to emotional associations—and VR tools and technology will evolve accordingly.

As a result, virtual reality may function as a link from the technological manifestations of humanity back to the world that technology has ostensibly supplanted. VR may transform our understanding of computers from severed heads to extensions of our whole selves. And in doing so, it may even offer a way to imagine ourselves, technology and all, as part of the natural world.

# Commentary: Satellites for a Developing World

*Satellites could provide universal access to the information economy.*

• • •

Russell Daggatt

Most people on this planet do not have access even to the most basic telephone service. To cite just one statistic, more than half the world's population lives more than two hours' travel time from the nearest telephone. The high cost of wire-line infrastructure has often kept telecommunications services from remote areas—as well as many not so remote areas.

Vast regions of the developing world are completely without telephone service. India has 860 million people but only about seven million telephone lines, virtually all of them clustered in a few large cities. Even in the U.S., despite its universal service principle, rural service is often woefully inadequate for the 21st century.

Where basic telephone service is available, the networks over which it is provided consist of 100-year-old technology—analog copper wire—that for the most part will never be upgraded to the digital, broadband capability required for the advanced network connections that have come to be known as the information highway.

As the rapid movement of information becomes increasingly essential to all those things we associate with a high standard of living—from education and health care to economic development and public services—there is a real danger that the quality of life will not improve, and may even decline, in areas that lack a digital broadband infrastructure.

There will be plenty of optical fiber, with its enormous capacity, linking countries and telephone companies. But fiber connections between individual offices and homes will be much rarer, even in heavily industrial regions.

As others have observed elsewhere in this book (see Chapter 2, "Wireless Networks," by George I. Zysman), wireless systems, whether they employ satellites, cellular transceivers or some combination of the two, can provide a way to extend the principle of universal service to underserved areas at low cost. Satellites and cellular systems are not interchangeable, however. To understand why, one must look ahead to what the developing world will need to participate in the future information-based economy.

The large fraction of the world that either has no access to basic telephone service or has access only through low-data-rate cellular systems needs a complementary technology that can inexpensively support broadband data and multimedia applications. In the developing world, these applications are not likely to be primarily for personal use but rather will be shared in such institutions as hospitals, schools, government offices and businesses.

In health care, for example, doctors and other caregivers can consult with specialists thousands of miles away, share medical records and images, relay critical information during epidemics, distribute globally the latest results of their research, expedite routing of medical supplies during disaster-relief efforts and give remote instruction in nutrition, sanitation and infant care.

With universal access to interactive broadband capabilities, information can flow freely between people, creating wider communities. In this context, satellite systems can complement terrestrial cellular systems in providing advanced broadband information infrastructures in areas where it would not be economically feasible to string wires.

Some of the advantages offered by satellites parallel those of land-based cellular systems. Satellites can serve vast areas at a cost that is indifferent to location. Satellite terminals can be deployed much more quickly and flexibly than cables can be laid. Moreover, because subscribers are not given exclusive use of satellite channels, the costs can be spread over many users all sharing the same resources.

But satellites can have some advantages over cellular systems. As the distances between users and the variability of the traffic increase, satellites' wider areas of coverage can make them more cost-effective. Satellites are also invulnerable to surface calamities, such as earthquakes, floods, fires and hurricanes, that cripple terrestrial communications systems.

Perhaps most important, satellite communications may help stem the large-scale migration of people from the countryside to cities and from the developing world to developed nations. Wire-line technologies just extend the industrial-age paradigm in which the economics of infrastructure drives people into overcrowded, overburdened urban congregations. Satellites can help people choose where they live and work based on such considerations as family, community and quality of life rather than access to infrastructure.

To the extent that the information revolution is based on optical fiber, the "information highway" metaphor is apt. Like a highway, or the railways before them, fiber is rigidly dedicated to a particular location. If a town is near the main line, it prospers; if it is a few miles distant, it dries up and blows away. Especially in the developing world, this model is becoming increasingly untenable. Moving information, instead of people, can create value and prosperity without consuming vast amounts of physical resources.

Advanced technologies have revolutionized the way people exchange and process information in urban areas of the U.S. and other developed nations. But a broader need is going unmet. Today the cost of bringing modern communications to poor and remote areas is so high that many of the world's citizens cannot participate in the global community. Yet the benefits of the information revolution should be extended to all, including those who do not live near centers of commerce or industry, who do not have ready access to doctors, hospitals, schools or libraries and who are at risk of being shunted aside. Satellite communications systems have the potential to alter the industrial paradigm positively and dramatically—and with it the lives of millions.

# PART II
# TRANSPORTATION

# High-Speed Rail: Another Golden Age?

*Neglected in North America but nurtured in Europe and Japan, high-speed rail systems are a critical complement to jets and cars.*

• • •

Tony R. Eastham

I looked at her with the uncomprehending adoration one feels for locomotives. . . . Against the fan of light her great bulk looms monstrous, a raving meteor of sound and mass.

That trains were once routinely described so breathlessly may come as a surprise. But the golden age of rail travel was arguably at its zenith when in 1935 columnist Christopher Morley wrote those words for the *Saturday Review of Literature.* Dozens of trains left New York City's Grand Central Station every day, bound for Chicago, Montreal, St. Louis and the like. Extensive networks of lines pervaded the continent and moved people, food stocks, primary resources and industrial products. Passenger trains such as the *Golden Arrow* in the U.K., the *Orient-Express* in Europe and the *Zephyr* in the U.S. came to exemplify not just speed, power and comfort but technological progress itself.

Of course, the six decades since that time have encompassed the advent of commercial air travel, as well as of interstate and international highway systems all over the world. Both these transportation modes have continued to develop and even now seem poised for further advances. Such developments prompt the question: What will be the role of rail travel in a world of large, subsonic transport aircraft and advanced automobiles running on "smart" highways?

The answer, in many advanced countries, is that trains will play a very important role indeed. In these regions, rail services have been dramatically enhanced through an evolution of systems and technologies in societies that have never relied on the automobile in quite the same way most in North America have. In many parts of Europe and Asia, trains, rather than airplanes, are now the preferred means of travel on routes of about 200 to 600 kilometers. Their use of fast and technologically advanced train systems began decades ago and may be supplemented in coming years by even more advanced magnetic-levitation (maglev) trains. Steel-wheel-on-steel-rail trains are now operating at speeds of up to 300 kilometers per hour, and maglev trains are being developed and tested for introduction at speeds of 400 to 500 kilometers per hour, perhaps within 10 years.

In North America, the implementation of high-speed rail has been frustratingly slow. Interurban and commuter rail services now account for less than 2 percent of passenger miles per year. Rail still moves substantial freight, but even here trucks have become the dominant carrier.

Nevertheless, there is growing recognition that mobility on the continent is being threatened by clogged freeways in metropolitan areas and by "winglock," the analogous condition at hub airports at peak times. Sustaining mobility and economic development will demand a more balanced combination of rail, air and road travel. Thus, a rail renaissance is being proposed. Millions of dollars have been spent on evaluations, systems design studies, route surveys and ridership assessments for heavily traveled corridors between cities several hundred kilometers apart.

Such studies have benefited from long records of experience in other places. In 1964 in Japan, for example, the famed Shinkansen (bullet train) opened between Tokyo and Osaka (see Figure 4.1). Over the years, speeds have climbed from 210 to 270 kilometers per hour, decreasing the trip time on the 553-kilometer Tokyo-to-Osaka run from four to 2.5 hours. Japan's Shinkansen network now covers 2,045 kilometers, from Morioka in northern Honshu to Hakata in Kyushu, and carries 275 million passengers every year. At the same time, technology development continues, led by Japan Railways Central, one of the country's regional railway companies. Among the projects under way is a Super Train for the Advanced Railway of the 21st century (STAR 21), whose prototype has achieved 425 kilometers per hour.

It is France, however, that has the fastest commercial train system in the world, the Train à Grande Vitesse (TGV). The TGV Atlantique has a maximum speed of 300 kilometers per hour. Paris forms the hub of a network that extends north to Lille and the Channel Tunnel, west to Tours and Le Mans, and south to Lyons. TGV trains also operate into Switzerland. In 1992 they began running in Spain between Madrid and Seville, and by 1998 they are to travel between Seoul and Pusan in Korea.

Germany also has a high-speed train, the InterCity Express (ICE), which now zips along at 250 kilometers per hour between Hannover and Würzburg and also between Mannheim and Stuttgart. Like those in Japan and France, this system operates on a dedicated right-of-way, maximizing passenger and public safety by eliminating road crossings and by using advanced control.

Sweden has adopted a somewhat different approach in its X2000, which achieves a top speed of 220 kilometers per hour on the 456-kilometer line from Stockholm to Göteborg and gets the most out of existing railroad infrastructure by actively tilting the passenger compartment relative to its wheeled undercarriage. The scheme avoids subjecting passengers to uncomfortable lateral forces while rounding curves at high speed. In Italy the ETR-450 tiltbody train provides similar service between Rome and Florence.

In North America, experience with high-speed rail has been limited for the most part to paper

studies and demonstrations of European technology to stimulate public interest. Many reports have evaluated the suitability of high-speed rail for such corridors as Pittsburgh–Philadelphia, Las Vegas–Los Angeles, San Francisco–Los Angeles–San Diego, Dallas–Houston–San Antonio, Miami–Orlando–Tampa and Toronto–Ottawa–Montreal. Nothing has been built, however, because the economics are projected to be marginal and because federal and state governments are reluctant to commit substantial funds.

Amtrak, the U.S. rail passenger carrier, does plan significant upgrades on its flagship Northeast Corridor routes, however. Part of this network, between Washington, D.C., and New York City, already runs at speeds of up to 200 kilometers per hour and carries more passengers than either of the competing air shuttles. In the immediate future, Amtrak plans to award a $700-million contract to purchase and maintain up to 26 high-speed train sets for use in the corridor; depending on budget negotiations in the U.S. Congress, the contract could be awarded later this autumn. The new trains will link Boston, New York City and Washington, D.C., with service of up to 225 kilometers per hour, once upgrading of the railroad infrastructure is finished, possibly in 1999. The trains are to be manufactured in the U.S., most likely as a joint venture with an offshore developer. Companies competing for the Amtrak contract are offering TGV, X2000 and ICE/Fiat technologies.

## Flying Low

Many of these promising applications for high-speed rail stem from the technology's evolutionary nature—most projects will keep costs down by making use of existing infrastructure. This is an advantage not shared by high-speed rail's revolutionary counterpart—maglev trains. Maglev is the generic term for a family of technologies in which a vehicle is suspended, guided and propelled by means of magnetic forces. With its need for an entirely new infrastructure, maglev is likely to find application primarily in a few heavily traveled corridors, where the potential revenue could justify the cost of building guideways from the ground up.

Mainly because of this obstacle, maglev has had a prolonged adolescence. The first conceptual outlines were published some 30 years ago by two physicists at Brookhaven National Laboratory on Long Island, N.Y. James R. Powell and Gordon Danby envisioned a 480-kilometer-per-hour (300-mile-per-hour) train suspended by superconducting magnet coils. Within a decade, however, virtually all the research and development shifted to Germany and Japan, which pursued different technical variations with substantial government and private funding.

With maglev, alternating-current electricity is fed to windings distributed along the guideway, creating a magnetic wave into which the vehicle's magnets are locked. Speed is controlled by varying the

**Figure 4.1 JAPANESE "BULLET" TRAIN reaches speeds of 275 kilometers per hour. This stretch, near Mount Fuji, is about 100 kilometers west and south of Tokyo, on the way to Osaka. The original line linked the two cities in 1964. Since then, the network has grown to cover some 2,045 kilometers, from Morioka in northern Honshu to Hakata, on the southern island of Kyushu.**

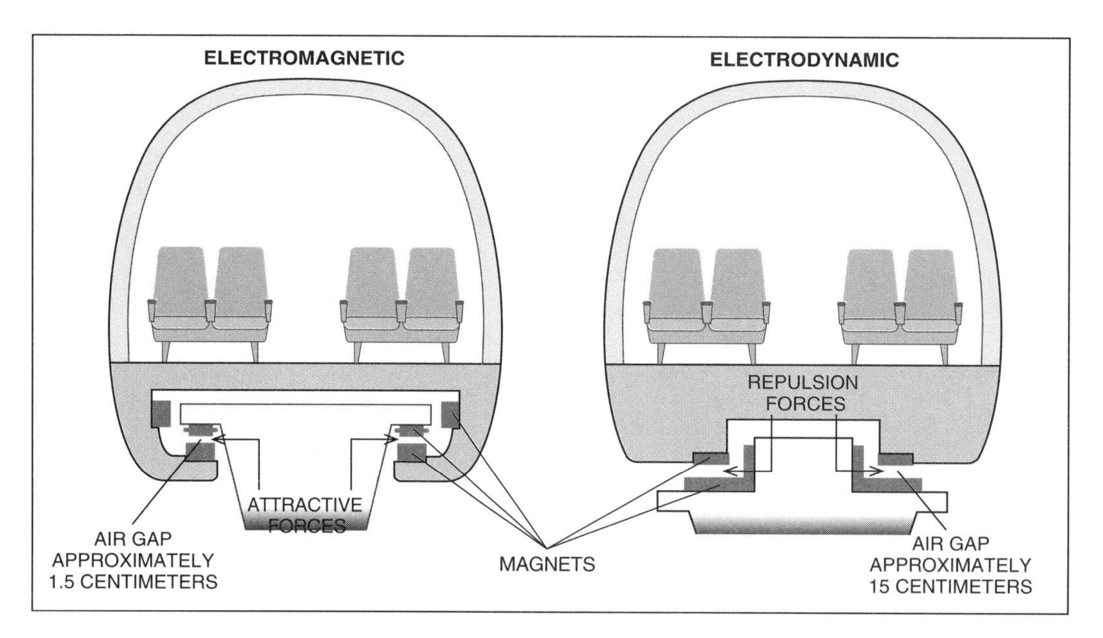


**Figure 4.2 HIGH-SPEED MAGLEV VEHICLES employ one of two kinds of suspension: electromagnetic (EMS) or electrodynamic (EDS). EMS relies on attraction between vehicle-mounted electromagnets and others on the underside of the guideway. In contrast, the electrodynamic system pushes the vehicle up above the guideway using repulsing magnets. EDS is based on superconducting magnets, creating a gap about 10 times greater than EMS is capable of producing. The greater gap allows for less precision in guideway construction tolerances. The ride quality of EDS vehicles is poorer than for EMS ones and requires more development.**

frequency of the electrical energy applied to the guideway windings. In effect, the vehicle's magnets and the windings in the guideway constitute a single synchronous electric motor, which provides linear rather than rotational motion.

There are two variations on this theme (see Figure 4.2). The so-called repulsion-mode electrodynamic system, proposed by Powell and Danby and pursued in Japan, uses superconductive magnets on board the vehicle to induce currents in conductive coils in the guideway. This interaction levitates the vehicle about 15 centimeters, as though it were a low-flying, guideway-based aircraft. Indeed, the Japanese vehicle achieves magnetic liftoff at about 100 kilometers per hour; at lower speeds, it rolls on wheels.

The other type, which has been developed in Germany, is the attraction-mode electromagnetic system. Conventional (nonsuperconducting) iron-core electromagnets carried by the vehicle are attracted upward toward ferromagnetic components attached to the underside of the guideway structure. This type of magnetic suspension is inherently unstable and needs precise control to maintain a clearance of about 1.5 centimeters between the vehicle's magnets and the guideway. One advantage, however, is that the vehicle remains levitated even when motionless and thus could be used for urban and commuter transit as well as for longer, high-speed routes. Indeed, the first operational maglev system was a low-speed shuttle installed in 1984 between the airport terminal and nearby railway station in Birmingham, England.

Japan's repulsion-mode system is being developed by the country's Railway Technical Research Institute in collaboration with a number of large engineering companies. A series of test vehicles included the ML-500R, which in 1979 achieved a speed of 517 kilometers per hour—a record for maglev—on a seven-kilometer test track near Miyazaki, on the island of Kyushu. Its successor, a prototype vehicle, will start test runs in 1997 on a 42.7-kilometer precommercial test and demonstration facility in Yamanashi prefecture near Tokyo.

This double-track guideway will allow essentially all aspects of an operational system to be tested, including full-size vehicles going through a tunnel at 500 kilometers per hour. Backers of the project are hopeful that a commercial version could be ready for deployment between Tokyo and Osaka by 2005.

In Germany the attraction-mode electromagnetic "Transrapid" maglev system has been under development by Magnetbahn GmbH since the late 1960s. Again, test vehicles led to the construction of a demonstration facility, at Emsland in the early 1980s. Its 31-kilometer, figure-eight-shaped guideway allows full-scale vehicles to run under conditions similar to operational ones. The preproduction vehicle TR-07 has been under evaluation for almost five years, regularly achieving speeds of from 400 to 450 kilometers per hour. The German government chose the technology for a new line linking Berlin and Hamburg. The route, to be built by about 2005, will be the centerpiece of a program to enhance east-west travel in the reunified Germany.

### U.S. Maglev: Suspended Animation

In the U.S., maglev development was abandoned after a brief period of research from the late 1960s to the mid-1970s at Ford Motor Company, the Stanford Research Institute and the Massachusetts Institute of Technology. The concept was rejuvenated in the late 1980s, however, and a government-sponsored National Maglev Initiative was launched in an attempt to apply some relevant technologies—cryogenics, power electronics, aerodynamics, control and vehicle dynamics—from the aerospace and related industries. The goal was a second-generation maglev system to meet the needs and conditions of North America.

In 1994 government funding ran out without spurring any sustained private-sector commitment. Four innovative maglev systems were designed. None were built, but the exercise generated several interesting ideas, including novel concepts for synchronous propulsion and a superconducting version of electromagnetic suspension with a large air gap between vehicle and track. By 1994 maglev R&D in North America had returned to its previous minimal state.

To some extent, maglev finds itself a victim of changing circumstances. Twenty or 25 years ago the technology was thought to be ideal for connecting densely populated areas up to 600 kilometers apart. Speeds of 450 to 500 kilometers per hour would make maglev competitive with air travel, it

## Rolling or Floating at 300 Kilometers per Hour

Advanced ground transportation has three categories: high speed, very high speed and magnetic levitation (maglev). High-speed systems, such as Amtrak's Northeast Corridor in the U.S., use the traditional steel-wheel-on-rail technology and can operate at top speeds ranging from 200 to 240 kilometers per hour (125 to 150 miles per hour). Very high speed systems are considered capable of reaching 350 kilometers per hour (218 mph), using enhanced wheel-on-rail technology. They are always electrically powered and require relatively straight route alignments to accommodate the higher speeds. The French Train à Grande Vitesse (TGV), Germany's InterCity Express (ICE) and Japan's Shinkansen (bullet train) are all examples of very high speed rail.

Maglev systems are quite different from traditional trains. They use electromagnetic forces to levitate, guide and propel train cars along a guideway at projected speeds of 320 to 500 kilometers per hour (200 to 310 mph). The German Transrapid and Japan's MLU—both noncommercial prototypes—are the only full-scale examples of high-speed maglev technology. A feature unique to maglev is the use of a synchronous motor that provides linear rather than rotational motion, with power supplied to magnet windings in the guideway.

*—John A. Harrison, Parsons Brinckerhoff Quade & Douglas, Inc.*

was reasoned, amid concerns about the cost and availability of oil-based fuels.

This argument largely depends on the maximum speed of steel-wheel-on-rail trains being significantly less than the speed of maglev; otherwise these conventional trains could in many cases fill the bill more economically. Two or three decades ago the practical speed limit for steel-wheel-on-rail was generally thought to be about 250 kilometers per hour. Yet, as noted, high-speed rail has developed to the point that operating speeds have reached 300 kilometers per hour. The achievement has followed from a better understanding of wheel-rail dynamics, aerodynamics and the transferring of high levels of electric power to a moving train from an overhead line. Even more impressively, wheeled trains have been tested at speeds of up to 520 kilometers per hour—three kilometers per hour faster than the maglev record. Although no one claims it would be feasible to run a passenger train at this speed, 350 kilometers per hour is now considered to be operationally workable.

Thus, maglev's speed and total-trip-time advantage is not what it once was; realistically, it would seem to be about 20 or 30 percent, in comparison with the best wheel-on-rail systems. It remains to be seen how many governments will find this margin compelling enough to commit to a fundamentally new transportation technology, for those few medium-range, heavily traveled routes where market share might be won from the airlines.

Of course, higher speeds would make maglev more attractive, and 500 kilometers per hour is not the final, upper limit by any means. But one of the main limiting factors at such speeds is aerodynamics. The power needed to overcome aerodynamic drag increases as the cube of speed; noise from aerodynamic sources increases as the sixth power. And the dynamic perturbations caused by trains passing one another or entering and exiting tunnels become increasingly severe at high speed. Such factors have led to proposals to run maglev trains in a fully or partially evacuated tube.

Years ago a study suggested that such a tunnel could link New York City, Los Angeles and perhaps other international cities as well through transoceanic links to provide the ultimate in global transportation. Top speed could be as high as 2,000 kilometers per hour, and a dipping and rising profile between stations would let gravity assist in propulsion and braking. Engineering considerations, such as the cost of building and maintaining such a tunnel, make the idea fanciful, to say the least.

Although it may be many decades before it will be possible to make reservations for the two-hour trip from New York to Los Angeles, important milestones are ahead for both maglev and more conventional, wheel-on-rail trains. The next decade should see the inauguration of the first moderately long-distance commercial maglev routes. High-speed rail, meanwhile, will be steadily enhanced in speed, comfort and passenger amenities. European services will become increasingly networked. At the same time, many more lines will be built in Asia, including the completion of a national network in Japan and new routes in Korea, Taiwan and China.

The U.S. is clearly a follower rather than a leader in the new high-speed rail technologies. But the country will find it necessary to rejuvenate its rail passenger routes, starting with the Northeast Corridor. Probably, some additional motivation will be needed, such as another oil crisis, or road and air congestion so bad that it interferes with economic growth. True, videoconferencing and other forms of telecommunications will lessen the need to travel and will save time and money. Nevertheless, there is no hard evidence yet that such communications facilities are slowing the growth of business travel. High-tech trains will come to North America—it is just a matter of time, need and a more favorable economic environment. Telecommunications may be the next best thing, but being there is best of all.

# The Automobile: Clean and Customized

*Built-in intelligence will let automobiles*
*tune themselves to their drivers*
*and cooperate to get through crowded traffic systems safely.*

• • •

Dieter Zetsche

Decades ago, as motor vehicles were becoming ubiquitous in many countries, images of drivers cruising open roads or moonlit parkways conjured the very essence of progress and autonomy. And for good reason: in developed countries, at least, the automobile was literally widening people's horizons, greatly expanding the area within which they could live, work and relax.

But as anyone who has ever endured a long and tedious traffic jam knows, automobiles must also operate as part of a huge, complex and sometimes unpredictable system. Through its arteries flow streams of vehicles driven by people with a variety of skill levels and all kinds of mental states. Steadily increasing numbers of vehicles have begun overloading this system in many urban areas, while contributing to an air pollution crisis in at least a few. Indeed, conservative estimates put the worldwide costs associated with accidents, wasted fuel and pollution at hundreds of billions of dollars a year.

So significant are these concerns that they have prompted the world's automobile makers to do something they never have before: join forces with one another and with a host of technical firms and research institutes to chart the course of automotive transportation in the next few decades. These ambitious programs, undertaken separately in Europe and the U.S., addressed safety and environmental and economic concerns; what was most notable about them, however, was their focus on traffic as a whole. The change of perspective portends a fundamental shift in the development of the automotive transportation system and, quite possibly, its most extensive metamorphosis since its beginnings more than a century ago.

For the first time, automobiles will be able to see, hear and communicate with one another and with the roadway itself. They will become sensitive to their drivers, warning of fatigue, distraction or exceeded speed limits. Gradually, vehicles will progress from being completely controlled by the driver to depending on the driver mainly for steering. In the more distant future, cars might even drive themselves on well-marked roads in good condition and in certain situations, such as when following other cars. And although the vast majority of them will continue to be powered by internal combustion engines, ultraefficient models will appear, along with growing numbers of hybrid-electric and fully electric vehicles.

In Europe, much of the conceptual work for this grand plan was carried out between 1986 and 1994 as part of a program called Prometheus (Program for European Traffic with Highest Efficiency and Unprecedented Safety). The project was a collaboration among 13 leading automobile manufacturers, some 50 electronic firms and distributors, and an equal number of research institutes. Both the U.S. and Japan have corresponding projects under way. The U.S. effort is called the Intelligent Vehicle and Highway System; Japan's program, which is concentrating on transmitting traffic information at intersections, is known as the Vehicle Information Communication System.

## Three Tiers

To make their huge undertaking manageable, Prometheus's engineers and scientists broke the task down into three fields of research, covering traffic management, cooperative driving and safety. Most technologies for the last two await development or introduction; however, pilot projects have already demonstrated most of the infrastructure for the traffic management system.

One of its cornerstones will be navigational assistance, from onboard computers receiving broadcast traffic reports. The driver begins by entering a destination into the vehicle's navigational computer. The computer chooses the fastest route, taking into account current traffic conditions, and guides the driver with verbal instructions over the vehicle's sound system and through a dashboard-mounted display screen. Variations of this system are now being introduced in the U.S., Japan and Europe.

In the European version, known as Dual-Mode Route Guidance, traffic reports will be updated continually, digitally encoded and broadcast over a special Traffic Message Channel of the Europe-Wide Radio Data System. Nor does the system limit itself to automobiles: if the message channel reveals that all the routes to a destination are jammed, and a convenient public-transportation alternative exists, the navigational computer will recommend it (as soon as this part of the system is available). Several automakers are now offering the navigational computers in their luxury models.

Along with most of the other systems being developed under Prometheus, the navigational computer can be considered a building block of a driverless automobile. Such a dramatic application, which is still at least several decades away, might initially be used at an airport or factory, shuttling cargo or equipment through fairly predictable routes and conditions. Encouraging results were achieved recently in a collaboration, called VITA II, among Daimler-Benz and several German universities. A Mercedes-Benz sedan was outfitted with 18 video cameras, which focused on the vehicle's surroundings. The car's position in its lane, traffic signs, obstacles and other traffic were all sensed and decoded, and a computer processed the information to drive the car in this realistic highway environment. During VITA II and its predecessor, VITA I, a total of about 5,000 kilometers were logged in test runs, mostly on German highways, at speeds of up to about 150 kilometers per hour. Developers are now considering ways of making the technology commercially viable.

## Cooperative Approach

Such intriguing possibilities notwithstanding, cars will, of course, continue to be driven by people for many years to come. Thus, accidents, road congestion and inclement weather will continue to pose hazards, to both vehicle occupants and the efficient flow of traffic. Prometheus's researchers concluded that the keys to lessening these dangers are information exchange and cooperation among vehicles, so they developed systems for these functions under a project named, appropriately enough, cooperative driving.

At the heart of their plan is a roadside information pool, which will gather information from regional traffic management centers and transmit it to automobiles from infrared or microwave beacons located along the road. In effect, the pool will be available to extend the driver's knowledge of traffic and road conditions far beyond his or her field of vision. For example, when a collision occurs, the vehicles involved are automatically located, and rescue measures initiated, by specially equipped mobile telephones.

To reduce the possibility of subsequent collisions, the emergency call will also warn approaching vehicles of the danger ahead—again, through the roadside pool. To help avoid accidents in the first place, the pool could rebuke drivers electronically for exceeding the speed limit by some margin. Most of the technology that would make all this possible was demonstrated during the Prometheus program and now awaits implementation.

Another feature bound to make driving safer, particularly for the easily distracted, is a new type of cruise control with far more intelligence and autonomy. By manipulating the brake and gas pedals, it will adjust the distance between a vehicle and the one in front of it "on the fly" for any combination of speed and road conditions. Incorporated into the vehicle's front end, the system will use radar or infrared beams to measure the distance to the car in front and will not require any special equipment or systems in the roadway.

Such traffic-wide approaches will be complemented, thanks to Prometheus's safe driving program, by improvements to the vehicles themselves. Many of these will help the driver see better through bad weather and in the dark. Already some carmakers are testing headlights that emit some ultraviolet radiation; besides reducing glare, the beams would be better reflected by pedestrians' clothing, lane markings and so on. Before long, some high-end automobiles will have infrared-vision systems and pulsed headlights that sense the scene in front of the vehicle without blinding other drivers.

The image will be brightly displayed, possibly with a heads-up system, which projects the scene so that the driver sees it as though it is floating some distance in front of the car. Optionally, the system could incorporate a range-control device that recommends—or even imposes—an appropriate speed after analyzing the driver's field of vision.

Sharper views of the road ahead can reduce but not eliminate close calls, so sensors, software and control strategies are in the works to help drivers maneuver or stop suddenly. Swerving, for instance, can be prevented by braking each wheel individually, with those most forward in the swerve braked more than the others. In addition, a number of automakers are experimenting with radar-based systems that would warn of an impending collision and possibly even apply the brakes. Eventually, it will become feasible to keep the car automatically within a lane under normal driving conditions.

While monitoring its surroundings, the car of the future will keep tabs on its driver as well. Fatigue and loss of concentration are leading causes of accidents, especially on long trips. But tired drivers give themselves away: their reaction time increases, and steering becomes erratic; eyelids begin to close; and the electrical resistance of the skin goes down. Once such signs are detected, an audible alarm can alert the driver it is time for a rest.

## Breathing Easier

If anything has become certain to automakers in recent years, it is that passenger safety is necessary but not sufficient; their products will have to be more environmentally benign as well. Practically every leading manufacturer is now committed to developing less polluting sources of motive power. Quite a few experimental vehicles powered by them have been tested, sometimes encouragingly and amid notable public interest. But for the next several decades, at least, almost all automobiles will continue to be powered by internal combustion engines, albeit cleaner and more efficient ones.

It takes little imagination to envision a market for advanced, ecologically friendly vehicles, distinguished by extreme fuel efficiency. These cars would be about twice as fuel efficient as today's thrifty cars; a liter of gasoline would propel them about 25 kilometers in mixed driving conditions (urban and highway, according to a standard German test procedure called DIN). One essential technology for these ultraefficient cars will be advanced, electronic motor-transmission management systems, which will increase efficiency through heightened sensitivity and interaction between the engine and the load on it (see Figure 5.1).

Of course, electric vehicles are the closest thing to a nonpolluting transportation medium. Their only adverse effects on the environment would be linked to the generation of electricity to charge their batteries and, possibly, to disposal of the batteries as well. Inadequate batteries are, in fact, the only significant obstacle to widespread use. An acceptable combination of energy and power densities, cost and service life still eludes developers, and consequently, electric vehicles will probably be confined to relatively small niches for at least the next 10 years.

The so-called hybrid electric has both a battery-powered electric motor and a small combustion engine, which are used either separately or together, depending on the driving situation. In some modes, the little engine runs continuously, efficiently and relatively cleanly, charging the batteries and greatly extending the vehicle's range. This advantage has prompted renewed interest in hybrids, with most large carmakers developing models to meet quotas for nonpolluting or low-polluting vehicles to take effect in some regions in the next five or six years.

Another method of cutting emissions significantly is to use fuels that cause less pollution. With respect to hydrocarbon emissions, methanol and liquefied or compressed natural gas burn more cleanly than does gasoline and have fueled experimental engines that have reached efficiencies comparable to those of conventional internal combustion engines. The cleanest fuel of all would be hydrogen, whose only significant combustion byproduct is steam. Derived from water, hydrogen is potentially abundant and versatile—it can be burned in a combustion engine or converted to electricity silently and without moving parts in a fuel cell, which could in turn power an electric motor.

Although both approaches look promising, the supply problems look much more intractable. At present, there is no way to produce the necessary amounts of hydrogen in an ecologically and economically acceptable manner. In Germany a liter of gasoline costs about US $1.10; to produce an amount of liquid hydrogen with comparable energy content costs at least US $2.00, depending on how the fuel is produced and transported. What is more, an infrastructure to distribute liquid hydrogen, comparable to today's network of gasoline stations, would require an impossibly large investment.

## Driving by Wire

Regardless of how they are powered, automobiles will in 20 or 30 years become increasingly attuned and even customized to their owners or drivers. Advanced manufacturing and "drive-by-wire" technologies will make possible a new generation of automobiles that can be reconfigured to suit not only different functions and needs but also a great array of strengths and weaknesses of different drivers.

At least one automaker has proposed basic vehicles whose elements would be drawn from a series of dozens or perhaps even hundreds of modules, foreshadowing a far greater variety of automobiles

**Figure 5.1 SLIPSTREAM over a coupe is manifested by smoke. Better fuel efficiency demands improvements in aerodynamics as well as in the engine and power train.**

**Figure 5.2 ROADSTER of the future nods to celebrated predecessors with 1950s-style headrests. This Mercedes-Benz concept car has a hard top and rear window that retract into the trunk at the push of a button.**

than we are used to seeing on roads today. These vehicular building blocks could be easily assembled and reassembled—at the time of purchase or indeed at any time during the automobile's useful life. On warm weekends or during a vacation, for example, the vehicle's functional frame might be turned into a sporty, open-air convertible (see Figure 5.2) or converted to a pickup to carry recreational equipment. The concept need not be limited to the chassis; an electric vehicle might have a charging-engine module that could be quickly installed, turning the car into a hybrid with greater range.

Customization could be taken to higher levels with the advent of drive-by-wire technology, which became standard in recent years in jet aircraft (in which it is known as fly-by-wire). With drive-by-wire, all the cables and mechanical connections that link, for instance, the steering wheel, pedals and shift lever to the axle, throttle and transmission will be replaced with electronics, including sophisticated control systems. The changeover to this technology will be an invisible revolution, requiring no new skills of the driver.

On the contrary, the technology will adapt to the driver's requirements and abilities, while opening up room for creative development. For example, drive-by-wire will permit stabilization of a vehicle in different configurations (convertible versus cargo-carrying pickup) and for different kinds of drivers: the driving enthusiast would get the response of a sports car, while less demanding drivers would get a smoother ride.

Within 30 years or so, innumerable automated vehicles, each constructed and finely tuned to its driver's needs and capabilities, could be making the roads far safer than they are today. Clearly, automotive development has not decelerated. In fact, it is just getting in gear.

# Evolution of the Commercial Airliner

*Advances in materials, jet engines and cockpit displays could translate into less expensive and safer air travel.*

• • •

Eugene E. Covert

Few people expected the dramatic changes in travel habits that followed when the jet-powered transport airplane was introduced into commercial passenger service in the late 1950s. Virtually free of vibration, this new airplane was quieter and more comfortable than the propeller-driven airplane that preceded it. Even more impressive was its ability to fly to Europe nonstop well above most storms, so that the ride remained extremely smooth. Shorter travel times and lower fares made the world more accessible for both the business and the casual traveler as well as for the transport of goods, from flowers to fish.

A future airliner may undergo equally dramatic changes. With the advent of advanced computer-aided design tools, airplanes may assume unusual shapes that offer higher performance and carry more passengers. The industry has already contemplated commercial airplanes in the shape of a flying wing, somewhat like the Stealth bomber (see Figure 6.1). A short, stublike body that holds the cockpit would protrude from the thick wing. But most of the passengers, as many as 800, would sit in the movie-theater-like space within the wing.

Other designs abound for airplanes with a wholly novel appearance. One futuristic airplane would be built with several fuselages interconnected by the wings and a series of struts. In addition, research engineers in the U.S. and Europe are investigating the feasibility of a supersonic airliner to serve as a successor to the Concorde.

A decision to go ahead with any of these plans may well hinge on the ability of the manufacturer to make the required multibillion-dollar investment in a highly uncertain economic climate. Indeed, the risks may be too high at this time. Boeing and some members of the Airbus Industrie consortium have recently issued results of a joint study that showed the lack of a market for huge airplanes that can carry 600 to 800 passengers. Such large airliners pose challenges not only to the aerospace engineering community but to airport operators as well: the longer wing span, for example, would require more space between passenger gates.

Even if none of these plans are ever realized, evolutionary improvements in many technologies will enable the next generation of conventional airplanes to run more economically than current airliners and to set new standards for safety and operating efficiency. In the immediate future, building airplanes with better materials and propulsion systems will continue to enhance performance. In the longer term, microscopic sensors and actuators, as well as computers distributed throughout the airplane, will yield a quiet revolution in the way engineers solve their problems. A "smart" airplane will give engineers control over many phenomena that they have long understood but could do little to change. The feedback and control mechanisms of tiny sensors and moving surfaces, called actuators, could reduce air resistance, or drag, or they could redistribute mechanical loads to increase the lifetime of a wing structure.

The information received from networks of sensors can be used to allow longer intervals between maintenance and will help a mechanic find the source of a problem quickly. Other sensors, when coupled with information from satellites, will be able to equip the aircrew with a better understanding of the whereabouts of nearby airplanes, a measure that will enhance safety.

**"Smart" engines, equipped with sensors and actuators, will adjust the air flowing through them to improve the efficiency and longevity of the propulsion system.**

**Pilots may one day react more quickly to unforeseen events by controlling cockpit displays with brain waves. Wright-Patterson Air Force Base has conducted research on controlling a flight simulator with brain waves.**

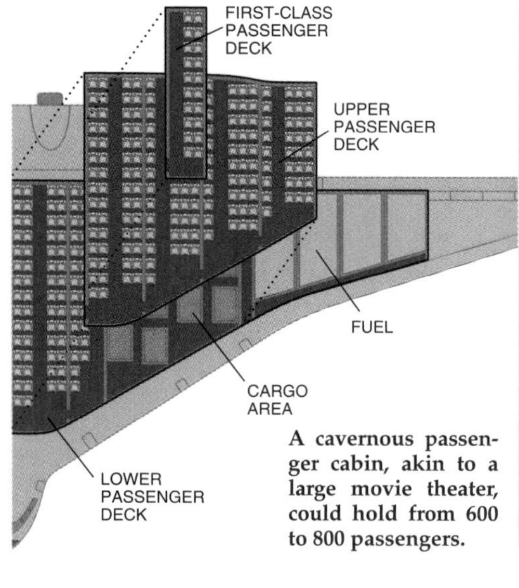


**A cavernous passenger cabin, akin to a large movie theater, could hold from 600 to 800 passengers.**

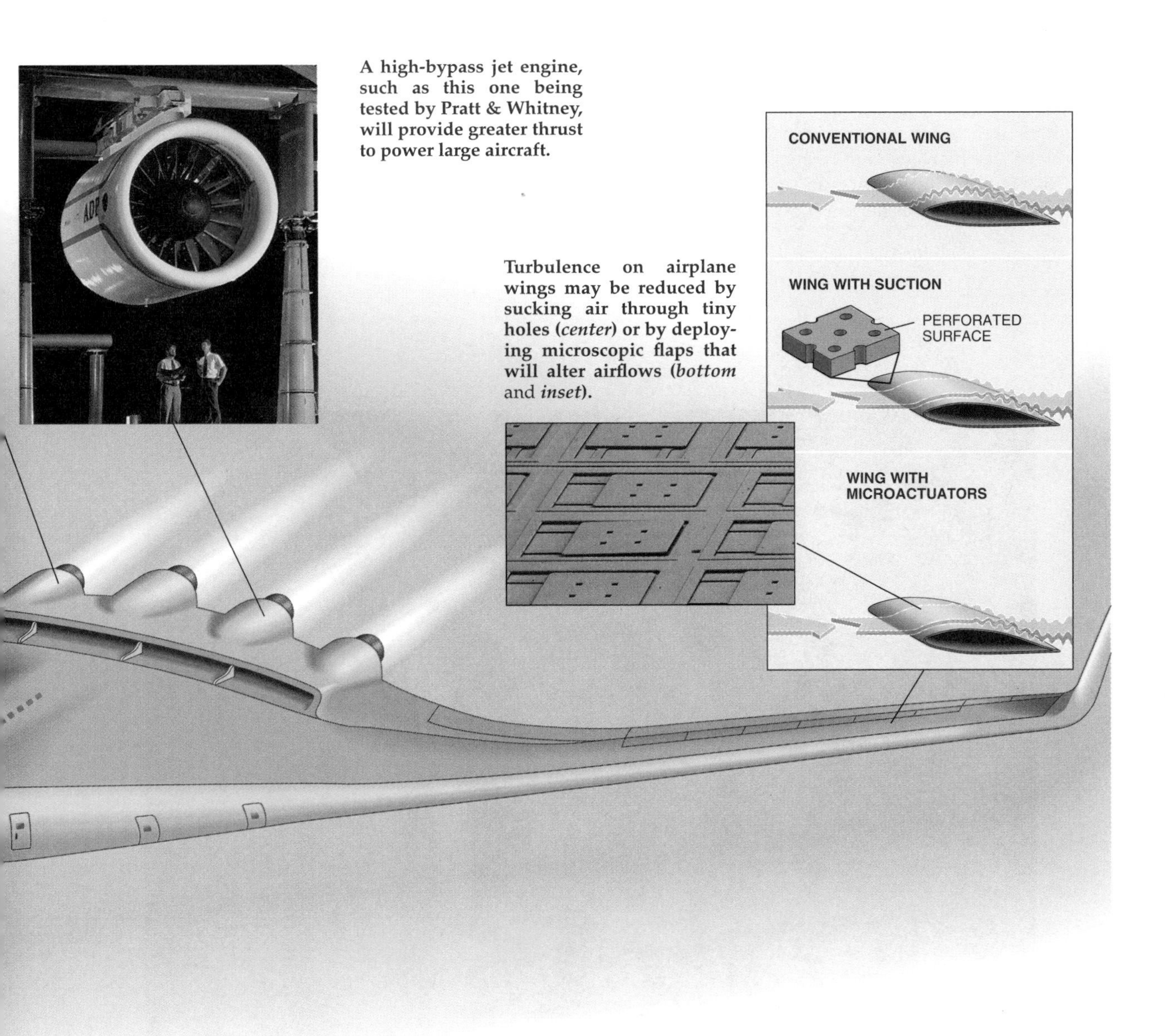


A high-bypass jet engine, such as this one being tested by Pratt & Whitney, will provide greater thrust to power large aircraft.

Turbulence on airplane wings may be reduced by sucking air through tiny holes (*center*) or by deploying microscopic flaps that will alter airflows (*bottom* and *inset*).

Figure 6.1 FLYING WING has been suggested as a design for a giant 600- to 800-passenger commercial airliner that would be built with multiple passenger decks. Besides assuming novel shapes, airplanes of the future may incorporate technologies ranging from advanced engines to cockpit controls manipulated by brain waves.

## Material Benefits

The smart airplane may take decades to evolve. Over the next 10 years, however, airlines will continue to push manufacturers to reduce the costs of operating a multimillion-dollar machine. Computer design software, with which engineers traditionally show how different parts and systems fit together, will help meet this goal. Large databases of airplane components will determine which parts are most reliable and which would offer the lowest cost of operation. Also over the next decade, structural designers will work with materials to reduce the weight of the fuselage and wings. One promising material is an aluminum-lithium alloy. It has a lower density and higher strength than other aluminum alloys, the reason Airbus Industrie incorporated it in the leading sections of the wings of the Airbus A330 and A340.

More widespread use of aluminum-lithium alloys is likely to be deferred until developers produce a material that can better withstand fracture. Alcoa is developing an aluminum-lithium metal that is said to improve fracture toughness by more than one third over an existing aluminum alloy. A study commissioned by Alcoa shows the material could save about 12 percent of the tail weight on large airplanes, or about 650 pounds. An attractive alternative is composites in which graphite fibers are placed into a matrix of organic polymers, making the material stronger per unit of weight. A designer can tailor materials to accommodate the different

**Figure 6.2 AEROSPACE COMPOSITE is manufactured in a plasma chamber at a plant owned by Textron Specialty Materials in Lowell, Mass. A worker extracts a semifinished sheet formed by spraying titanium onto a bed of silicon carbide fibers.**

loads carried on a structure by varying the number of layers of fibers and the direction in which the fibers are placed (see Figure 6.2).

For a decade or more, military and civilian airplanes have incorporated some composites in structures that bear light loads. The cost of the materials—and the imperfect understanding of how they fail—has delayed their broader introduction into commercial airplanes. Yet these hurdles are slowly being overcome. Composites now make up 9 percent of the structural weight of the Boeing 777, which went into service this June, about triple the amount in the 757 and 767 airplanes.

Because most composites are cured in a hot oven under pressure, they are said to be "thermoset." But they cannot be heated again and reworked into another shape. This drawback may be overcome by "thermoplastic" polymers that can be reshaped to eliminate a flaw introduced in manufacturing.

The longevity of both aluminum and composite materials may be enhanced by embedding in them tiny sensors to measure local levels of strain. This information could assess the remaining life of the structure as well as provide feedback to adjust the movable flight-control surfaces—the flaps or ailerons—to reduce mechanical loading.

## Better Jet Engines

Composites made of metal and ceramics may enhance engine performance. For example, an engine made with silicon carbide fibers embedded in a titanium matrix could run at higher temperatures and thus produce the same amount of thrust with less fuel. The engine would also be lighter and longer lived.

Engineers have also devised other means to boost efficiency. In today's jet engine, air is drawn in by a fan, pressurized and then mixed with fuel and burned before the hot gases stream out the back as exhaust. The efficiency of this propulsion system, its ability to convert heat into thrust by burning the hot air-fuel mixture, is enhanced by allowing some of the air that enters the engine to move through a duct that bypasses the engine "core" where the hot gases are burned. Such a "high bypass" engine, as it is called, also generates less noise.

In modern gas-turbine engines the bypass ducts handle six or seven times the amount of air passing through the core. Some experimental engines have raised the bypass air to 20 times that passing through the core. These ultrahigh bypass engines will come with a price, however. They will undoubtedly increase the size and weight of the fan and other engine components in addition to augmenting drag.

The efficiency of a gas-turbine engine also depends on the shape of the rotating blades in the compressor, where air is pressurized, and in the turbine that powers the compressor. The Massachusetts Institute of Technology's Gas Turbine Laboratory is developing a smart engine laden with sensors and actuators that could alter the shape of the blades while in flight to improve performance.

A more efficient engine combined with lower drag will improve the economics of airplane operation even more. One source of drag, a result of air passing over the wings, presents a particularly difficult problem. Aerodynamicists study the behavior of thin strata of air near the surface of an airplane, which are collectively known as the boundary layer. In a laminar boundary layer, airflow stays smooth. The layers of air pass by one another as if they were playing cards in a deck, one sliding easily over the other. But in large passenger airplanes, these regular currents become swirling eddies of turbulence, inducing substantial amounts of drag.

Experiments by the National Aeronautics and Space Administration using a modified Boeing 757 have shown that drag can be reduced by literally sucking the turbulence away. To achieve this objective, a portion of a wing is covered with many tiny holes that are connected to a suction pump. The action of the pump draws air into the holes, thereby smoothing out the flows in the boundary layer. The amount of turbulence can be detected by minute sensors on the wing, which signal how much suction is needed. Some measures could reduce turbulence with tiny actuators: in one approach, thousands of microscopic, movable flaps hinged to the wing would rise from the surface base when the airflows turn turbulent (see Chapter 12, "Engineering Microscopic Machines," by Kaigham J. Gabriel).

## Situation Awareness

Pilots will perform better if they have tools that give them a better understanding of where they are and where they are going. "Situation awareness," as it is called, describes pilots' ability to integrate diverse instrument readings for speed, altitude, location, weather, direction of flight and the presence of nearby aircraft. An engineer must decide how best to display information to the aircrew. Most of the time the pilot needs to see only that the needles on the dials point in the proper direction. In an emergency, however, a pilot must quickly assess the danger and determine alternative courses of action.

Today's airplane instrumentation is capable of disseminating an enormous flood of data—too much, in fact, for the pilot to comprehend in the few critical seconds available. Software and processors are required that can screen and then show only data relevant to the situation. Before such systems can be programmed, a new discipline is needed—one that might be called cognitive engineering—that would combine studies in how humans absorb information and how these perceptions are related to activities that take place in the brain.

Cognitive engineering may prove most effective when combined with advanced computer displays, which project the most important information onto a visor mounted on a pilot's helmet. One day a pilot may respond to data on a screen as fast as the mind can think. Research has progressed on the use of brain waves to change the position of a cursor or to manipulate other cockpit controls.

The airplane of the year 2050 may or may not resemble the airplanes one flies in today. But advances in aviation technology and an understanding of pilots' cognitive processes will make air transportation safer and accessible to more travelers.

# 21st-Century Spacecraft

*A fleet of cheap, miniaturized spacecraft*
*may revive the stalled Space Age,*
*exploring the myriad tiny bodies of the solar system.*

• • •

Freeman J. Dyson

The first question we have to answer in discussing the future of space activities is, "What went wrong?" The Space Age, which began with a flourish of trumpets about 40 years ago, was supposed to lead humanity onward and upward to a glorious future of cosmic expansion. Instead it became, like the Age of Nuclear Power, a symbol of exaggerated expectations and broken promises. We now find ourselves living in the Information Age, in which the technologies transforming our lives are not rocketry and astronautics but microchips and software. Space activities play only a subsidiary part, supporting the satellite links that serve as alternatives to fiber-optic channels on the ground. The Space Age fizzled because the grand dreams turned out to be too expensive. From now on, space technology will thrive when it is applied to practical purposes, not when it is pursued as an end in itself.

One of those practical purposes is scientific research. Here, too, there has been some confusion between ends and means. Some space science projects grew so big and expensive that they were driven by politics and bureaucratic momentum rather than by science. Missions on a grand scale, such as the *Voyager* flybys of the outer planets and the *Hubble Space Telescope* observations of distant galaxies, have returned a wealth of knowledge and have brought political glory to their sponsors. But within the National Aeronautics and Space Administration, as in the world outside, winds of change are blowing. Billion-dollar missions are no longer in style. Funding in the future will be dicey. The best chances of flying will go to ventures that are small and cheap.

In 1995 I spent some weeks at the Jet Propulsion Laboratory (JPL) in Pasadena, Calif. JPL, which built and operated the *Voyager* spacecraft, is the most independent and the most imaginative part of NASA. I was interested in two innovative planetary missions proposed by JPL, the Pluto Fast Fly-by and the Kuiper Express. The *Pluto Fast Fly-by* would complete *Voyager*'s exploration of the outer solar system by taking high-resolution pictures in many wave bands of Pluto and its large satellite Charon. The *Kuiper Express* would similarly explore the Kuiper Belt, the swarm of smaller objects orbiting beyond Neptune whose existence was predicted about 40 years ago by the astronomer Gerard P. Kuiper.

## The Big Shrink at JPL

Both mission concepts were based on a radical shrinkage of the instruments carried by the *Voyager* probes. I held in my hands the prototype package of instruments for the new spacecraft. It weighs five kilograms but does the same job as the *Voyager* instruments, which weighed more than 200 kilograms (see Figure 7.1). All the hardware components—optical, mechanical, structural and electronic—have been drastically reduced in size while providing significantly improved sensitivity.

Daniel S. Goldin, the administrator of NASA, encouraged JPL to devise these missions to carry on the exploration of the outer solar system using spacecraft radically cheaper than *Voyager*. The two Voyager missions, which began in 1977, each cost about $1 billion. The JPL designers estimated that their proposal for the Pluto Fast Fly-by mission would cost $500 million. Goldin told them, in effect, "Sorry, but that is not what I had in mind."

The *Pluto Fast Fly-by* failed because it did not depart radically enough from the concept and technology of *Voyager*. It still carried a heavy thermoelectric generator, using the radioactive decay of plutonium 238 as the source of energy. It still relied on massive chemical rockets to give it speed for the long voyage from here to Pluto. It was new wine in an old bottle. The scientific instruments were drastically miniaturized, but the rest of the spacecraft was not reduced in proportion.

After Goldin rejected the *Pluto Fast Fly-by*, the designers worked out a revised version of the mission, now called the Pluto Express. The Pluto Express shrinks the entire spacecraft, but it still uses chemical rockets and a plutonium power supply. It is new wine in a half-new bottle. The estimated cost is now about $300 million for two independently launched spacecraft. The designers hope to launch them around the year 2003.

Meanwhile the *Kuiper Express* picks up where the *Pluto Express* stopped. The *Kuiper Express* is new wine in a new bottle. It is the first radically new planetary spacecraft since the first Pioneer missions in the late 1950s (see Figure 7.2). The *Kuiper Express* dispenses with chemical rockets. Its propellant is xenon, an inert gas that can conveniently be carried as a supercritical liquid, as dense as water, without refrigeration. The prototype engine was undergoing endurance tests in a tank at JPL when I visited. It must run reliably for 18 months without loss of performance before it can be seriously considered for an operational mission.

**Figure 7.1 MINIATURIZED INSTRUMENTS will help reduce the size and cost of future spacecraft. Patricia M. Beauchamp of the Jet Propulsion Laboratory, co-leader of the study team for the *Kuiper Express*, holds an example of a downsized instrument package that could be used for such a mission. The 1970s-technology *Voyager* instruments, seen at the right, are less sensitive but weigh 40 times as much.**

The power source for the *Kuiper Express* is a pair of large but lightweight solar panels. The panels are expansive enough to provide power for instrument operations and communication with the earth even in the dim sunlight in the Kuiper Belt. No plutonium generator is needed. Some JPL experts reject the idea of using solar energy to transmit signals home from such a great distance and doubt whether the huge, flimsy solar collectors required for the job will ever be practical. But the designers of the *Kuiper Express* decided to jettison the last heavy piece of *Voyager* hardware so their bird can fly fast and free.

**Figure 7.2 *KUIPER EXPRESS* spacecraft represents a sharp break from current space exploration technology. In place of bulky chemical rockets, a lightweight ion engine propels the craft; the emerging ions emit a blue light. Power for the engine comes from two huge but lightweight solar panels. Radically shrunken hardware improves spacecraft performance without sacrificing scientific capability. The *Kuiper Express* would examine the unexplored, comet-like bodies orbiting beyond Neptune. One such object looms ahead while the receding crescent of Neptune (*shown exaggerated in size*) reflects off the spacecraft in this artistic rendering of a design proposal.**

The *Kuiper Express* is a daring venture, breaking ground in many directions. It demands new technology and a fresh style of management. It may fail, like the *Pluto Fast Fly-by,* because its architects may be forced by political constraints to make too many compromises and to rely too much on old concepts. But solar-electric propulsion has opened the door to a new generation of cost-effective small spacecraft, taking advantage of the enormous progress over the past 30 years in miniaturizing instruments and computers. If the *Kuiper Express* fails to fly, some other, bolder mission will succeed. In space science, as in politics, the collapse of the old order opens possibilities for adventurous spirits.

Spaceflight will certainly change radically between now and the end of the next century. Nevertheless, physical technology is far enough advanced that we can make plausible guesses about where it will be in 100 years. Our guesses will be wrong in detail, but we know enough about the laws of physics to set firm limits regarding what we can and cannot do. The future of biotechnology is uncertain in a different way. We cannot even guess where its limits lie. I believe the 21st century will be the Age of Biotechnology and that biotechnology will transform the shape of spacecraft as profoundly as it will transform the patterns of human life. I will describe here only the physical machinery whose form and performance can be predicted with some confidence.

## A New Way to Fly

Many systems of propulsion have been proposed for the spacecraft of the future. Five systems—nuclear-electric propulsion, solar-electric propulsion, laser propulsion, solar sails and electromagnetic ram accelerators—offer great technical promise. Each has

**Figure 7.3 ION ENGINE (*at arrow*) may be the propulsion technology of choice in the next century—it is cheap, simple and efficient. This prototype is undergoing endurance tests in a huge vacuum chamber at the Lewis Research Center in Cleveland. The inset photograph shows a side view of the engine during a test firing. Electricity passes through xenon gas in the engine; the resulting electrically charged atoms, or ions, are pulled out of the engine, creating the glowing, propulsive jet. The modest but persistent thrust from ion engines could accelerate spacecraft to high velocities, speeding the exploration of the solar system.**

particular advantages for certain missions. Forced to choose one, I pick solar-electric as the most promising.

Solar-electric propulsion accelerates a spacecraft by means of a low-thrust ion jet (see Figure 7.3). Sunshine falling on solar cells generates electricity, which is used to ionize and accelerate a nonreactive gas, such as xenon. The positively charged ions are pulled out of the engine, forming a jet that impels the craft forward. In a chemical rocket, the fuel provides both energy and momentum, whereas in solar-electric propulsion, the sources of energy (sunlight) and momentum (the ion jet) are separate. I declare solar-electric propulsion to be the winner in space because it allows us to push as far in the directions of speed, efficiency and economy as the laws of physics allow. I see solar-electric as a cheap, general-purpose engine for moving payloads all over the solar system.

This flexibility does not mean that other propulsion systems will not be necessary. To launch from the surface of the earth into space, we will still need chemical rockets or some more efficient high-thrust launcher. Nuclear-electric operates much like solar-electric but eliminates the dependence on sunlight for electricity. Laser propulsion draws its energy from a high-power laser located on the earth's surface; it is well suited to a rapid-fire launch schedule. Solar sails, driven by the pressure of sunlight, get off to a slow start but require no fuel. Ram accelerators provide an inexpensive launch system for rugged bulk-freight payloads that can stand an acceleration of several thousand gravities. But for long-range, high-speed hauling, either of freight or of passengers, solar-electric seems the propulsion system of choice.

There are two physical limits to the performance of solar-electric spacecraft. One is set by the thrust-to-weight ratio of ion-jet engines, the other by the power-to-weight ratio of surfaces that can collect the energy of sunlight. The thrust-to-weight ratio constrains the acceleration of the spacecraft and is most important for short missions. For instance, if a vessel can accelerate at one centimeter per second squared (roughly one thousandth of earth gravity), it will be able to reach a speed of 26 kilometers per second in a month's time. Covering relatively small distances quickly requires a fast start and hence a high thrust-to-weight ratio.

The power-to-weight ratio constrains the maximum velocity of the spacecraft and is most important for long journeys. The thinner the surface of the solar collector, the greater the power for a given weight. The laws of physics put the limit of thinness at about one gram per square meter. This is less than one thousandth of the weight of the solar panels on the proposed *Kuiper Express*, so a solar-electric spacecraft could in principle surpass the performance of the *Kuiper Express* by a factor of 1,000.

The *Kuiper Express* would use its solar-electric engines for propulsion near the earth and then coast the rest of the way to the edge of the solar system. A spacecraft built around thin-film collectors could use its solar engine anywhere in the solar system, even in the Kuiper Belt, where sunlight is 1,000 times weaker than it is here. In the neighborhood of the earth, such a craft could achieve speeds of several hundred kilometers per second, fast enough to reach Mars or Venus or Mercury in much less than a month. A diversified system of solar-electric spacecraft would make the entire solar system about as accessible for commerce or for exploration as the surface of the earth was in the age of steamships.

## Setting Anchor in Space

In the next century, as in the last, everything will depend on making the passage cheap. Even if thin-film solar panels can be made inexpensively, their size and flimsiness will make them difficult to deploy. The problem of deployment will be much greater for large payloads. Spacecraft carrying people, for instance, probably cannot weigh less than one ton. A one-ton craft that provides high performance and good economy in the use of propellant will need solar collectors roughly two acres in area—a little larger than a football field. Whether such voyages make sense will depend on the motivation of the passengers.

I, like the majority of scientists, am more interested in unmanned missions. For robotic probes, the advantages of solar-electric propulsion can be realized more conveniently by shrinking the overall weight from tons to kilograms. The process of miniaturization, which led from *Voyager* to the *Kuiper Express*, can be carried much further. The typical solar-electric spacecraft of the future will probably weigh a few kilograms and carry a solar panel 10 to 20 meters in diameter. Such spacecraft would be ideal for scientific exploration and probably also for most commercial and military ventures. The main business of spacecraft, whether they are scientific, commercial or military, is information handling, which we already know how to do economically using machines that weigh less than a kilogram. Solar-electric propulsion will give such missions a flexibility that chemical rockets lack. Spacecraft using solar-electric propulsion may wander freely around the solar system, altering their trajectories to follow the changing needs of science.

Where will our fleets of microspacecraft be going 100 years from now? To ask this question today is like asking the Wright brothers in 1905 to predict where airplanes would be going in 1995. The simple answer to both questions is, "Everywhere." One important, often overlooked fact about the solar system is that the bulk of the real estate is not on planets. The planets contain most of the mass outside the sun, but most of the surface area is on smaller objects: satellites, asteroids and comets. Except for the larger satellites and a couple of asteroids, all this area is totally unexplored. Our spacecraft will mostly be going to places that nobody has yet seen.

Nature has kindly arranged the solar system so that most of the potential destinations are smaller bodies whose gravity is weak. Such places are directly accessible to low-thrust, solar-electric spacecraft. Landing spots for 21st-century spacecraft will be easier to find than deep harbors were for 19th-century steamships.

# Commentary: Why Go Anywhere?

*Millions of people could be liberated from their vehicles.*

• • •

Robert Cervero

In a decade or two, travel by automobile in some advanced countries may very well involve the kind of technology and intelligence gathering once reserved for tactical warfare. Onboard navigational aids, fed by satellite tracking systems, will give directions in soothing digital voices. In big cities, roadside screens will flash messages about distant traffic jams and alternative routes. Computerized control and guidance devices embedded underneath heavily trafficked corridors will allow appropriately equipped cars and trucks to race along almost bumper to bumper. Special debit cards will let motorists enter tollways without stopping, park downtown without fumbling for change, and hop on trains and people movers by swiping their cards through electronic turnstiles. Some experts even foresee customized tractor trailer–style "car-buses" that carry up to 20,000 cars per hour per freeway lane—10 times the current capacity.

This constellation of new-age technologies is only a part of the master plans drawn up in recent years in Europe and the U.S. For example, Detroit's "Big Three" automakers and the U.S. Department of Transportation have together spent or pledged several billion dollars through the remainder of this decade on R&D and commercialization of the Intelligent Transportation System. Although its goals of enhanced efficiency, comfort and safety are unimpeachable, its inevitable costs—spiraling fuel consumption, air pollution, suburban sprawl and urban decay—are sobering.

One sensible and compelling alternative would be to reduce the need to travel in the first place. Tools and techniques that could drastically cut back on commuting already exist; appropriately designed communities, for example, could put most destinations within walking (or bicycle-riding) distance, and telecommunications, computers and other technologies could let many people work from their homes or from facilities nearby. The potential savings in time, energy, natural resources and psychological wear-and-tear could be enormous.

Toward this end, a small but influential group of "new urbanist" planners and architects, led by Miami-based Andres Duany and Californian Peter Calthorpe, is seeking to re-create the pedestrian-friendly towns of yesteryear. Places such as Princeton, N.J., Annapolis, Md., and Savannah, Ga., are their paragons. These small cities have lively central cores within walking distance of most residents, prominent civic spaces, narrow, tree-lined streets laid out in grids and varying mixtures of housing and main-street shops.

**TOWN CENTER provides civic focus for Kentlands, a planned community in southern Maryland. Convergence of streets creates more routes from the center, distributing traffic. Another tenet of the "new urbanist" credo is close spacing of homes, allowing residents to walk to shops, parks and community centers and to socialize with one another more easily.**

Most notably, these are places where one is less inclined to drive and more likely to walk or bicycle to a nearby video store or delicatessen. In contrast to contemporary master-planned suburbs where people are confined to cars and homes, these "neotraditional" communities would allow people of all ages and walks of life to come into daily contact. Such mingling, new urbanists believe, will promote social cohesion and build stronger bonds between people and place. Critics charge that such humanist design principles smack of social engineering and physical determinism; of course, one could level similar charges against the federally subsidized interstate highways and home mortgages that nurtured the automobile industry and suburban sprawl in the postwar era.

Because of the tepid real estate market and skepticism among lenders and developers, few neotraditional towns have actually been built. Perhaps the best examples to date are Seaside, a compact, upscale resort community on Florida's Panhandle, and Kentlands, a Georgetown look-alike just outside Gaithersburg, Md. (see above photo).

By being less dependent on vehicles, life in these communities is more energy-efficient. Research by Reid Ewing of Florida Atlantic University shows that households in the sprawling suburbs of Palm Beach County log two thirds more vehicle hours per person than comparable households in more compact, traditional communities.

Self-contained towns in which people live, work and shop would be even more efficient. Nearly a century ago, to relieve London from overcrowding, the planner-architect Sir Ebenezer Howard first advanced the idea of building self-sufficient satellite communities separated by greenbelts and connected by railways. The British government embraced Howard's ideas, building two dozen such communities between 1946 and 1991, including settlements such as Milton Keynes and Redditch that were platted on open fields.

In Britain, most of these relatively young communities, called new towns, are highly self-contained—around two thirds of their workforce resides locally. In new towns outside Stockholm and Paris, on the other hand, most workers commute from outside the community. More than half do so efficiently, however, by rail. In comparison, most American new towns, such as Columbia, Md., and Reston, Va., remain predominantly bedroom communities, populated by drive-alone commuters. Other than remote retirement enclaves, America's most self-contained places (with low transportation-fuel consumption rates per capita) are its densest urban centers, Manhattan and San Francisco.

Advances in telecommunications and changes in the ways people live and work are now starting to bring some of this self-sufficiency to other regions. The number of contract workers, self-employed entrepreneurs and cottage industries is on the rise, and computers, multimedia devices and satellite communications are increasingly within the reach of average consumers. The developments will foster rapid growth in home-based enterprises in coming years.

Distributed workplaces will also take the form of neighborhood telecenters, equipped with videoconferencing, on-line data-search capabilities, facsimile transmission and voice mail, allowing suburbanites to walk or bike to their jobs a few days a week and work at home the other days. In Orange County and Sacramento, Calif., several developers have included home offices and neighborhood telecenters in new mixed-use developments. Neighborhood coffee shops and corner cafés have also opened, becoming watering holes where home-based workers socialize and "network." Today's cyberkids, reared on the Internet and interactive media, are apt to be even more receptive to the idea of working at home when they reach adulthood.

In recent years, technology has allowed a fortunate minority to live wherever it wants. Many of these people—software specialists, independent consultants, writers and the like—are choosing fairly out-of-the-way spots, like Sante Fe, N.M., central Colorado and Peterborough, N.H. Now other towns are picking up on the trend. In an effort to provide local residents with jobs and to keep people from leaving, Oberlin, Kan., has built a telecommunications post in its town center. Steamboat Springs, Colo., bills itself as a communications haven, actively luring professionals who work at home. Advanced technologies—such as satellite communications systems, the proposed information highway and very high speed trains—could spawn a constellation of self-contained urban villages across North America. Residents of these places would lose less of their time in transit.

The past 150 years has been a self-perpetuating cycle of urban transportation advances and decentralization. New transportation technologies have stretched the envelope of urban development, raising per capita fuel consumption, consuming farmlands and open space, and dirtying air basins. White flight to the freeway-laced suburbs and exurbs has left many inner cities in a state of near collapse and divided by race and class. America's vast highway system, of course, has just been a means to an end; rising incomes, residential preferences, prejudices and other social and economic factors motivated families to flee urban centers. By failing to pass on the true social costs to motorists, however, we encourage excessive auto travel and subsidize sprawl. The so-called Intelligent Transportation System stands to worsen this state of affairs by orders of magnitude.

The difference between advancing these costly transportation technologies as opposed to designing new kinds of sustainable communities is the difference between automobility and accessibility. Enhancing automobility—the ability to get from place to place in the convenience of one's own car—is and has been the dominant paradigm guiding transportation investments throughout this century.

Accessibility, in contrast, is about creating places that reduce the need to travel and, in so doing, help to conserve resources, protect the environment and promote social justice. Have technology, will travel. Have sustainable communities, will prosper.

# PART III
# MEDICINE

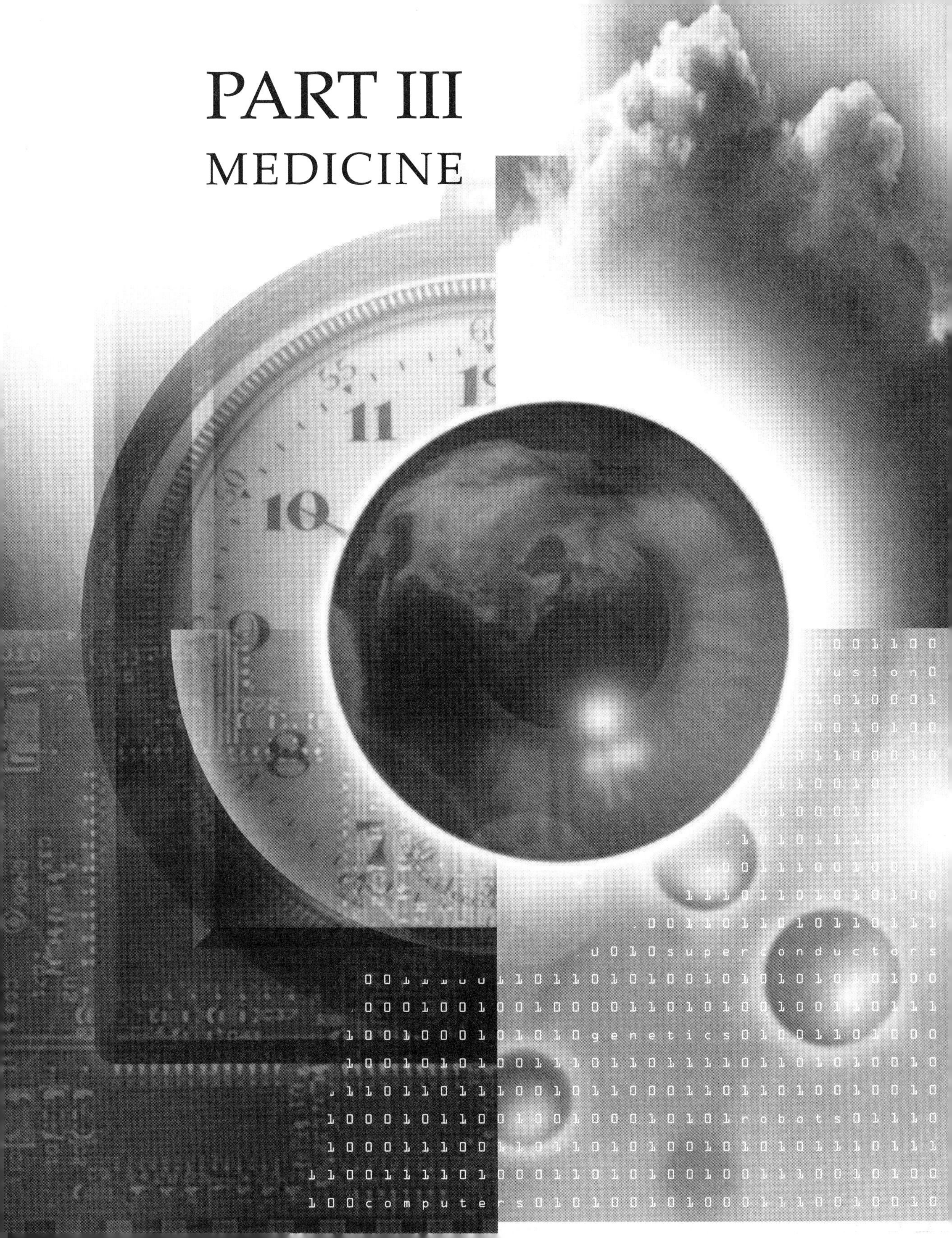

# Gene Therapy

*Several hundred patients have already received treatment. In the next century the procedure will be commonplace.*

• • •

W. French Anderson

On September 14, 1990, Ashanti DeSilva, barely four years old, became the first patient to undergo federally approved gene therapy. Little Ashanti was suffering from severe combined immunodeficiency (SCID) because she had inherited a defective gene from each parent. The gene in question normally gives rise to an enzyme, adenosine deaminase, needed for proper function of the immune system. Without this critical enzyme, Ashanti's immune system shut down, leaving her vulnerable to a host of infections.

The treatment Ashanti received was carried out at the National Institutes of Health by a team composed of R. Michael Blaese, Kenneth W. Culver, myself and several others. We removed white blood cells of the immune system from her body, inserted normal copies of the defective gene into the collection and then returned the treated cells to her circulation. The experiment went well. After Ashanti received four infusions over four months, her condition improved. With the help of occasional follow-up treatments, she has now been transformed from a quarantined little girl, who was always sick and left the house only to visit her doctor, into a healthy, vibrant nine-year-old who loves life and does everything (see Figure 8.1).

The new technology of gene therapy promises to revolutionize medicine in the next century. Over the course of history, there have been three great leaps in our ability to prevent and treat diseases. When society began taking such public health measures as establishing sanitation systems, vast numbers of people were protected from devastating infections. Next, surgery with anesthesia enabled doctors for the first time actually to cure an illness. (If someone has an inflamed appendix, for example, removal of the organ solves the problem for life.) Introduction of vaccines and antibiotics ushered in the third revolution, making it easy to prevent or correct many diseases spread by microbes.

Gene therapy will constitute a fourth revolution because delivery of selected genes into a patient's cells can potentially cure or ease the vast majority of disorders, including many that have so far resisted treatment. If this notion seems surprising, consider that almost every illness arises in part because one or more genes are not functioning properly. Genes give rise to proteins (the main

**Figure 8.1** ASHANTI DESILVA, the first recipient of federally approved gene therapy, was treated for a lethal inherited condition called severe combined immunodeficiency when she was four years old. With help from occasional booster treatments, she has grown into a healthy, active nine-year-old.

workers of cells), and defective genes can yield disease when they cause cells to make the wrong amount of a protein or an aberrant form of it.

More than 4,000 conditions, such as SCID and cystic fibrosis, are caused by inborn damage to a single gene. Many other ailments—the scourges of cancer, heart disease, AIDS, arthritis and senility, for example—result to an extent from impairment of one or more genes involved in the body's defenses. These defenses, all of which require genetically specified proteins, involve not only the immune system but also the body's mechanisms for maintaining itself. For example, liver cells manufacture proteins that help clear cholesterol from the blood. If an imperfection in the gene for this protein leads to a reduction in the abundance or efficiency of the protein, the result may be high cholesterol levels, atherosclerosis and heart disease.

Today understanding of the precise genetic bases for many diseases is still sketchy, but that knowledge will increase enormously in the next few decades. By the year 2000, scientists working on the Human Genome Project should have determined the chromosomal location of, and deciphered parts of the DNA code in, more than 99 percent of active human genes. And research aimed at uncovering the function of each gene will be progressing rapidly. Such information should make it possible to identify the genes that malfunction in various diseases.

## How Gene Therapy Works

To deliver corrective genes to damaged cells, researchers have developed several methods for transporting genetic material. The most effective technique employs modified viruses as such carriers. Viruses are useful in part because they are naturally able to penetrate cells, inserting the genetic material they contain into their new host. Before viruses can be used in therapy, however, genes coding for proteins that viruses use to reproduce and to cause disease must be removed. When those genes are replaced with a corrective gene, one has a delivery system that is identical to the original virus on the outside and can transport useful genes into cells but cannot cause illness.

Physicians can apply gene therapy in either of two ways. As was done in Ashanti's case, they may insert a healthy copy of a gene into the patient's cells in order to compensate for a defective gene. (The genes do not always have to be fit into a patient's chromosomes to be helpful; they simply have to survive and give rise to therapeutic levels of the specified protein.) Physicians may also introduce a purposely altered gene in order to give a cell a new property. For example, several groups are investigating such a treatment for patients infected with the human immunodeficiency virus (HIV), the cause of AIDS. In this therapy, copies of a gene that interferes with replication of HIV are inserted into the patient's blood cells, where they can potentially halt progress of the disease. Eventually, physicians might even deliver genes to prevent certain medical conditions. Instead of waiting for a woman who is susceptible to breast cancer to become ill, for example, doctors might provide her with protective genes while she is still healthy.

For the next decade, gene delivery is most likely to be performed only on somatic cells, which consist of all cell types except sperm, eggs and their precursors. Genetic alteration of somatic cells affects only the patient undergoing treatment. In theory, gene therapy could also be applied to the reproductive, or germ, cells. Modification of these cells would affect all descendants of the original patient. Many aspects of modern life, such as tobacco smoking and exposure to radiation, may well alter the genetic makeup of subsequent generations inadvertently, but deliberate application of germ-cell gene therapy would open a Pandora's box of ethical concerns that few investigators are willing to confront until much more is known about somatic-cell gene therapy.

## Recent Progress

Researchers have developed several approaches to gene therapy on somatic cells. The most commonly used technique, known as ex vivo (or "outside the living body") therapy, is the kind we used for Ashanti: cells with defective genes are removed from the patient and given normal copies of the affected DNA before being returned to the body. This therapy has generally targeted blood cells because many genetic defects alter the functioning of one type of these cells or another. But blood cells have limited life spans. Thus, corrected cells slowly disappear, and periodic treatments are typically required.

Future efforts will most likely target the stem cells of the bone marrow: the immature cells that give rise to the full array of blood cells in the circulation and that replenish blood cells as needed. Stem cells are ideal targets for gene therapy because they appear to be immortal: they survive as long as the patient lives. Hence, they constitute a permanent reservoir for an inserted gene.

Although scientists have been able to obtain stem cells from human bone marrow, they are having difficulty getting genes into these elusive cells as well as inducing the cells to produce many new blood cells in the body. Advances are being made, however. Just this year a team led by Donald B. Kohn at Children's Hospital in Los Angeles announced that three newborns afflicted with SCID who were treated by inserting genes into their stem cells are now thriving two-year-olds. Their blood cells are producing the critical enzyme they lacked at birth. Because young infants grow rapidly, their stem cells are quite active. In older patients, though, stem cells produce new blood cells more slowly. Fortunately, the problem is tractable. Investigators in many laboratories are making good progress toward isolating the chemicals the body itself uses to trigger stem cell division. I expect that by the beginning of the 21st century, the stem cell problem will be resolved.

A second, reasonably well developed method of somatic-cell gene therapy is called in situ (or "in position") treatment. In this procedure, carriers bearing corrective genes are introduced directly into the tissue where the genes are needed. This procedure makes sense when a condition is localized, but it cannot correct systemic disorders.

In situ treatment is being explored for several diseases. In the case of cystic fibrosis, which impairs the lungs, workers have introduced gene carriers containing healthy copies of the cystic fibrosis gene into the lining of the bronchial tubes. As a first step toward treating muscular dystrophy, other researchers have injected a gene directly into muscle tissue in animals to investigate the possibility of reengineering the body to make normal muscle proteins. Several teams have inserted "suicide" carriers into tumors; these carriers contain a gene intended to make cancer cells commit suicide when treated with certain chemotherapy drugs.

In situ therapy is still hampered by a lack of safe and effective ways for implanting corrected genes into various organs. Further, as is true of ex vivo therapy, the genes do not always yield good quantities of the encoded proteins. Also, the altered cells are rarely immortal—and so the useful genes perish when the cells that house them die. Finally, in both ex vivo and in situ therapy, once genes enter cells, they can insert themselves randomly into the DNA of chromosomes. Such insertion can be harmless; at times, however, it may have serious consequences. If, for example, the corrective genes disrupt tumor suppressor genes, which normally protect the body against tumors, cancer may result.

## Better Methods to Come

A third class of treatments, known as in vivo (or "in the living body") therapy, does not exist today, but it is the therapy of the future. Physicians will simply inject gene carriers into the bloodstream, in much the way many drugs are administered now (see Figure 8.2). Once in the body, the carriers will

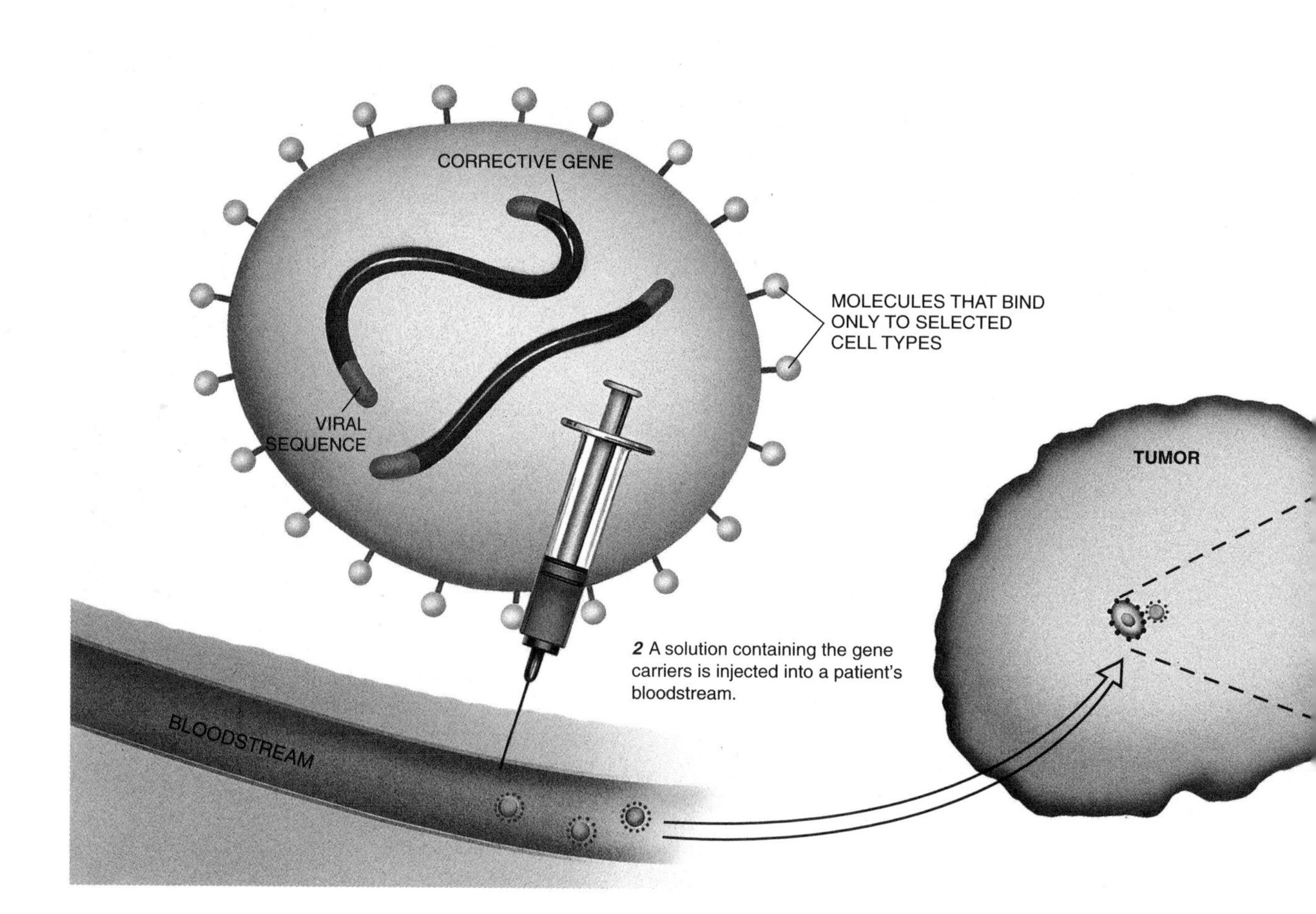

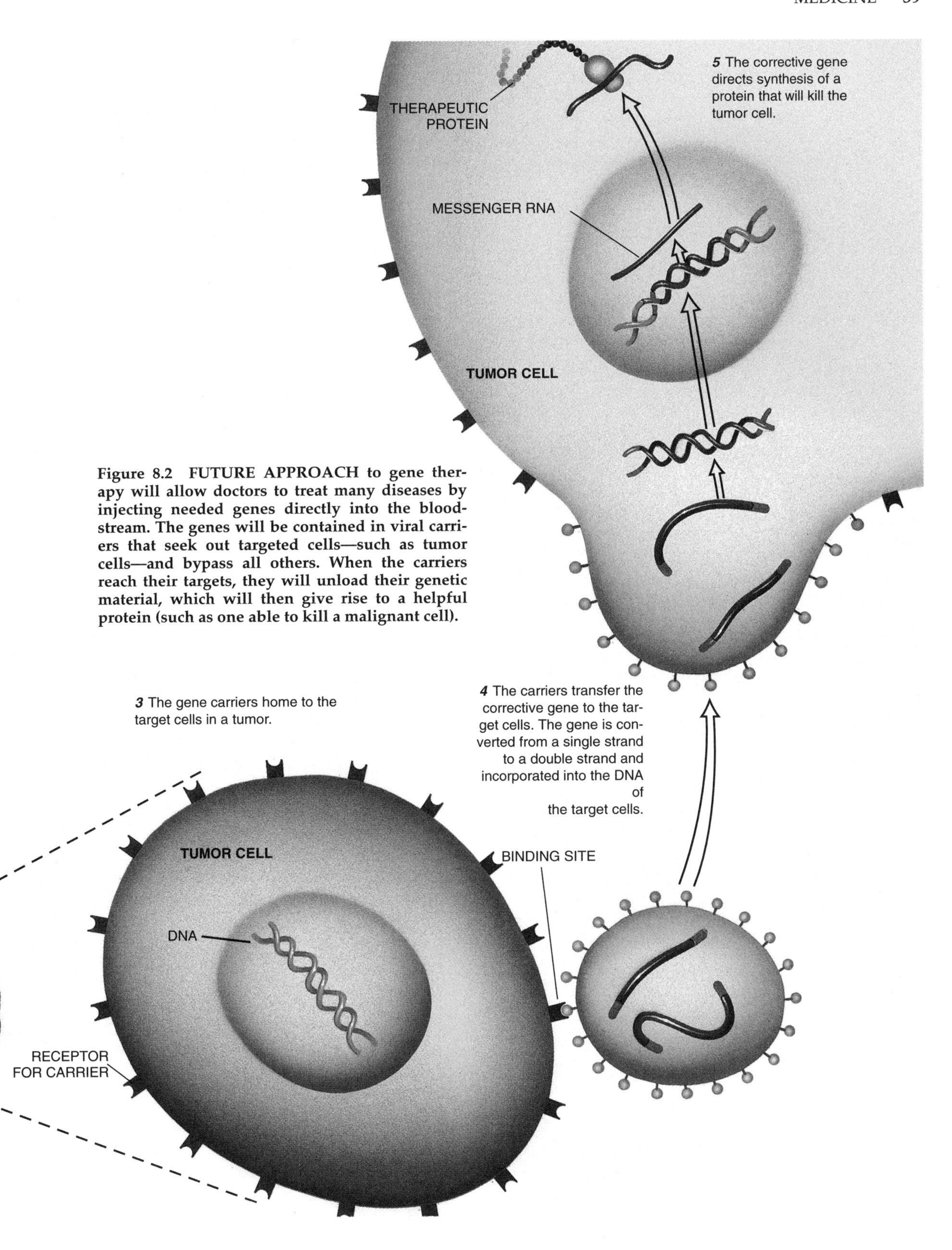


**Figure 8.2 FUTURE APPROACH to gene therapy will allow doctors to treat many diseases by injecting needed genes directly into the bloodstream. The genes will be contained in viral carriers that seek out targeted cells—such as tumor cells—and bypass all others. When the carriers reach their targets, they will unload their genetic material, which will then give rise to a helpful protein (such as one able to kill a malignant cell).**

## Diseases Being Treated in Clinical Trials of Gene Therapy

- Cancer (melanoma, renal cell, ovarian, neuroblastoma, brain, head and neck, lung, liver, breast, colon, prostate, mesothelioma, leukemia, lymphoma, multiple myeloma)
- SCID
- Cystic fibrosis
- Gaucher's disease
- Familial hypercholesterolemia
- Hemophilia
- Purine nucleoside phosphorylase deficiency
- Alpha-1 antitrypsin deficiency
- Fanconi's anemia
- Hunter's syndrome
- Chronic granulomatous disease
- Rheumatoid arthritis
- Peripheral vascular disease
- AIDS

find their target cells (while ignoring other cell types) and transfer their genetic information efficiently and safely.

Good progress has been made toward ensuring that vectors will home to specific cell types. But workers have had less success creating delivery units that can efficiently insert their genetic payload into the target cells or elude a patient's immune system and other defenses long enough to reach their destination. The ongoing concern about randomly inserted DNA segments remains a problem as well. Despite these hurdles, I am optimistic that by 2000 many solutions will have been found, and early versions of injectable vectors that target specific cells will be in clinical trials.

Aside from the technical challenges that will have to be overcome before gene therapy of any kind will be routine, there is the problem of cost. For the next several years, early trials will continue to be expensive and to be carried out only in major medical centers. But in contrast to some procedures, such as heart transplantation, that will always be expensive, gene therapy should ultimately become simpler and less costly. Indeed, within 20 years, I expect that gene therapy will be used regularly to ameliorate—and even cure—many ailments.

No discussion of gene therapy is complete without mention of its ethical implications. When we have the technical ability to provide a gene for correcting a lethal illness, we also have the ability to provide a gene for less noble purposes. There is thus a real danger that our society could slip into a new era of eugenics. It is one thing to give a normal existence to a sick individual; it is another to attempt to "improve" on normal—whatever "normal" means. And the situation will be even more dangerous when we begin to alter germ cells. Then misguided or malevolent attempts to alter the genetic composition of humans could cause problems for generations.

Our society went into the age of nuclear energy blindly, and we went into the age of DDT and other pesticides blindly. But we cannot afford to go into the age of genetic engineering blindly. Instead we must move into this exciting new era with an awareness that gene therapy can be used for evil as well as for good. As we reap the benefits of this technology, we must remember its pitfalls and remain vigilant.

# Artificial Organs

*Engineering artificial tissue is the natural successor to treatments for injury and disease. But the engineers will be the body's own cells.*

• • •

Robert Langer and Joseph P. Vacanti

In the third-century legend of Saints Cosmos and Damian, the leg of a recently deceased Moorish servant is transplanted onto a Roman cleric whose own limb has just been amputated. The cleric's life hangs in the balance, but the transplant takes, and the cleric lives. The miraculous cure is attributed to the intervention of the saintly brothers, both physicians, who were martyred in A.D. 295. What was considered miraculous in one era may become merely remarkable in another. Surgeons have been performing reimplantation of severed appendages for almost three decades now, and transplants of organs such as the heart, liver and kidney are common—so common, in fact, that the main obstacle to transplantation lies not in surgical technique but in an ever worsening shortage of the donated organs themselves.

In the next three decades, medical science will move beyond the practice of transplantation and into the era of fabrication. The idea is to make organs, rather than simply to move them. Advances in cell biology and plastic manufacture have already enabled researchers to construct artificial tissues that look and function like their natural counterparts. Genetic engineering may produce universal donor cells—cells that do not provoke rejection by the immune system—for use in these engineered tissues. "Bridging" technologies may serve as intermediate steps before such fabrication becomes commonplace. Transplantation of organs from animals, for example, may help alleviate the problem of organ shortage. Several approaches under investigation involve either breeding animals whose tissues will be immunologically accepted in humans or developing drugs to allow the acceptance of these tissues. Alternatively, microelectronics may help bridge the gap between the new technologies and the old. The results will bring radical changes in the treatment of a host of devastating conditions.

Millions of Americans suffer organ and tissue loss every year from accidents, birth defects and diseases such as cancer. In the last half of this century, innovative drugs, surgical procedures and medical devices have greatly improved the care of these patients. Immunosuppressive drugs such as cyclosporine and tacrolimus (Prograf) prevent rejection of transplanted tissue; minimally invasive surgical techniques such as laparoscopy have reduced trauma; and dialysis and heart-lung machines sustain patients whose conditions would otherwise be fatal.

Yet these treatments are imperfect and often impair the quality of life. The control of diabetes

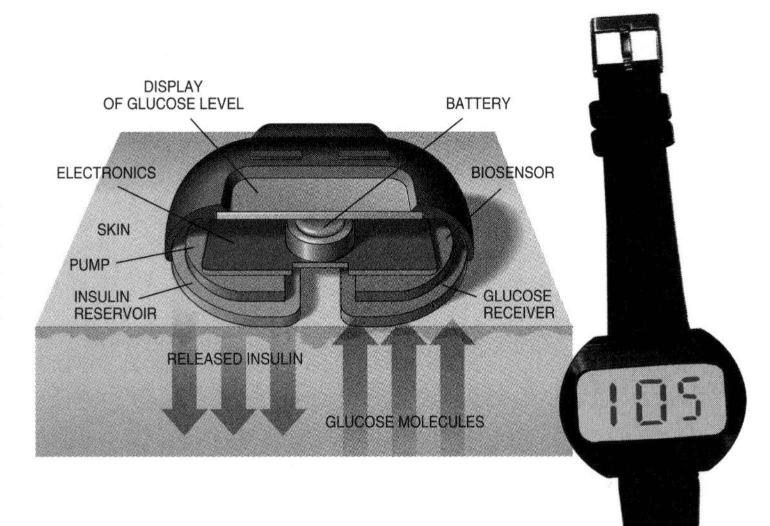


**Figure 9.1 INSULIN-DELIVERY SYSTEM on the drawing boards for diabetics would be worn like a watch. Electric fields or low-frequency ultrasound would temporarily increase the permeability of the skin, allowing glucose molecules to be withdrawn from the bloodstream. A sensor would pass insulin back through the skin in proportion to the amount of glucose detected. Glucose-sensing systems, something like the "wristwatch"prototype shown, are expected to start clinical trials in 1996; insulin-delivery technology is at an earlier stage.**

with insulin shots, for example, is only partly successful. Injection of the hormone insulin once or several times a day helps the cells of diabetics to take up the sugar glucose (a critical source of energy) from the blood. But the appropriate insulin dosage for each patient may vary widely from day to day and even hour to hour. Often amounts cannot be determined precisely enough to maintain blood sugar levels in the normal range and thus to prevent complications of the disease later in life—conditions such as blindness, kidney failure and heart disease.

Innovative research in biosensor design and drug delivery may someday make insulin injections obsolete. In many diabetics, the disease is caused by the destruction in the pancreas of so-called islet tissue, which produces insulin. In other people, the pancreas makes insulin, but not enough to meet the body's demands. It is possible to envision a device that would function like the pancreas, continuously monitoring glucose levels and secreting the appropriate amount of insulin in response. The device could be implanted or worn externally.

## An Artificial Pancreas

Much of the technology for an external glucose sensor that might be worn like a watch already exists (see Figure 9.1). Recent studies at the Massachusetts Institute of Technology, the University of California at San Francisco and elsewhere have shown that the permeability of the skin can temporarily be increased by electric fields or low-frequency ultrasonic waves, allowing molecules such as glucose to be drawn from the body. The amount of glucose extracted in this way can be measured by reaction with an enzyme such as glucose oxidase, or light sensors could detect the absorbance of glucose in the blood.

These sensing devices could be coupled via microprocessors to a power unit that would pass insulin through the skin and into the bloodstream by the same means that the sugar was drawn out. The instrument would release insulin in proportion to the amount of glucose detected.

An implantable device made of a semipermeable plastic could also be made. The implant, which could be inserted at any of several different sites in the body, would have the form of a matrix carrying reservoirs of insulin and glucose oxidase. As a patient's glucose level rose, the sugar would diffuse into the matrix and react with the enzyme, generating a breakdown product that is acidic. The increase in acidity would alter either the permeability of the plastic or the solubility of the hormone stored within it, resulting in a release of insulin proportional to the rise in glucose. Such an implant could last a lifetime, but its stores of glucose oxidase and insulin would have to be replenished.

The ideal implant would be one made of healthy islet cells themselves. That is the rationale behind islet cell transplantation [see "Treating Diabetes with Transplanted Cells," by Paul E. Lacy; SCIENTIFIC AMERICAN, July 1995]. Investigators are working on methods to improve the survival of the implants, but the problem of supply remains. As is the case with all transplantable organs, the demand for human pancreas tissue far outstrips the availability. Consequently, researchers are exploring ways to use islets from animals. They are also studying ways to create islet tissue, not quite from scratch, but from cells taken from the patient, a close relative or a bank of universal donor cells. The cells could be multiplied outside the body and then returned to the patient.

## Spinning Plastic into Tissue

Many strategies in the field of tissue engineering depend on the manipulation of ultrapure, biodegradable plastics or polymers suitable as substrates for cell culture and implantation. These polymers possess both considerable mechanical strength and a high surface-to-volume ratio. Many are descendants of the degradable sutures introduced two decades ago. Using computer-aided design and manufacturing methods, researchers will shape the plastics into intricate scaffolding beds that mimic the structure of specific tissues and even organs. The scaffolds will be treated with compounds that help cells adhere and multiply, then "seeded" with cells (see Figure 9.2). As the cells divide and assemble, the plastic degrades. Finally, only coherent tissue remains. The new, permanent tissue will then be implanted in the patient.

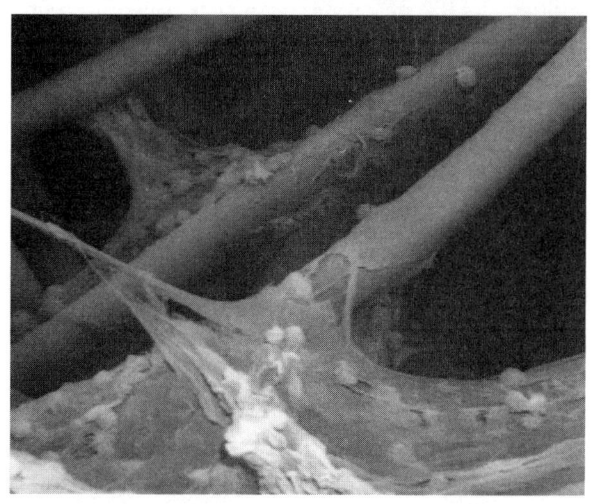

This approach has already been demonstrated in animals, most recently when our group engineered artificial heart valves in lambs from cells derived from the animals' blood vessels [see "Have a Heart" and "Science and the Citizen," SCIENTIFIC AMERICAN, June 1995]. During the past several years, human skin grown on polymer substrates

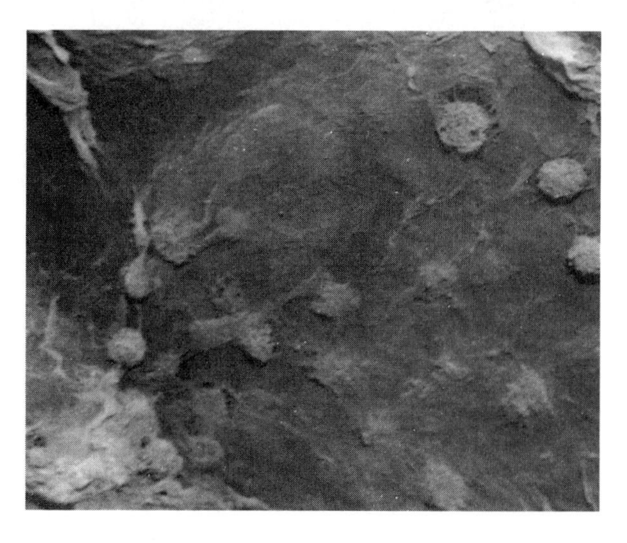

**Figure 9.2 SCAFFOLDING has been constructed to provide a template for formation of new tissue (shown enlarged, from top to bottom, ×200, ×500, ×1,000). The biodegradable plastic has been seeded with cells, which divide and assemble until they cover most of the structure. Eventually, the plastic degrades, leaving only tissue.**

has been grafted onto burn patients and the foot ulcers of diabetic patients, with some success. The epidermal layer of the skin may be rejected in certain cases, but the development of universal-donor epidermal cells will eliminate that problem.

Eventually, whole organs such as kidneys and livers will be designed, fabricated and transferred to patients. Although it may seem unlikely that a fully functional organ could grow from a few cells on a polymer frame, our research with heart valves suggests that cells are remarkably adept at organizing the regeneration of their tissue of origin. They are able to communicate in three-dimensional culture using the same extracellular signals that guide the development of organs in utero. We have good reason to believe that, given the appropriate initial conditions, the cells themselves will carry out the subtler details of organ reconstruction. Surgeons will need only to orchestrate the organs' connections with patients' nerves, blood vessels and lymph channels.

Similarly, engineered structural tissue will replace the plastic and metal prostheses used today to repair damage to bones and joints. These living implants will merge seamlessly with the surrounding tissue, eliminating problems such as infection and loosening at the joint that plague contemporary prostheses. Complex, customized shapes such as noses and ears can be generated in polymer constructs by computer-aided contour mapping and the loading of cartilage cells onto the constructs; indeed, these forms have been made and implanted in animals in our laboratories. Other structural tissues, ranging from urethral tubes to breast tissue, can be fabricated according to the same principle. After mastectomy, cells that are grown on biodegradable polymers would be able to provide a completely natural replacement for the breast.

Ultimately, tissue engineering will produce complex body parts such as hands and arms (see Figure 9.3). The structure of these parts can already be duplicated in polymer scaffolding, and most of the relevant tissue types—muscle, bone, cartilage, tendon, ligaments and skin—grow readily in culture. A mechanical bioreactor system could be designed to provide nutrients, exchange gases, remove waste and modulate temperature while the tissues mature. The only remaining obstacle to such an accomplishment is the resistance of nervous tissue to regeneration. So far no one has succeeded in growing human

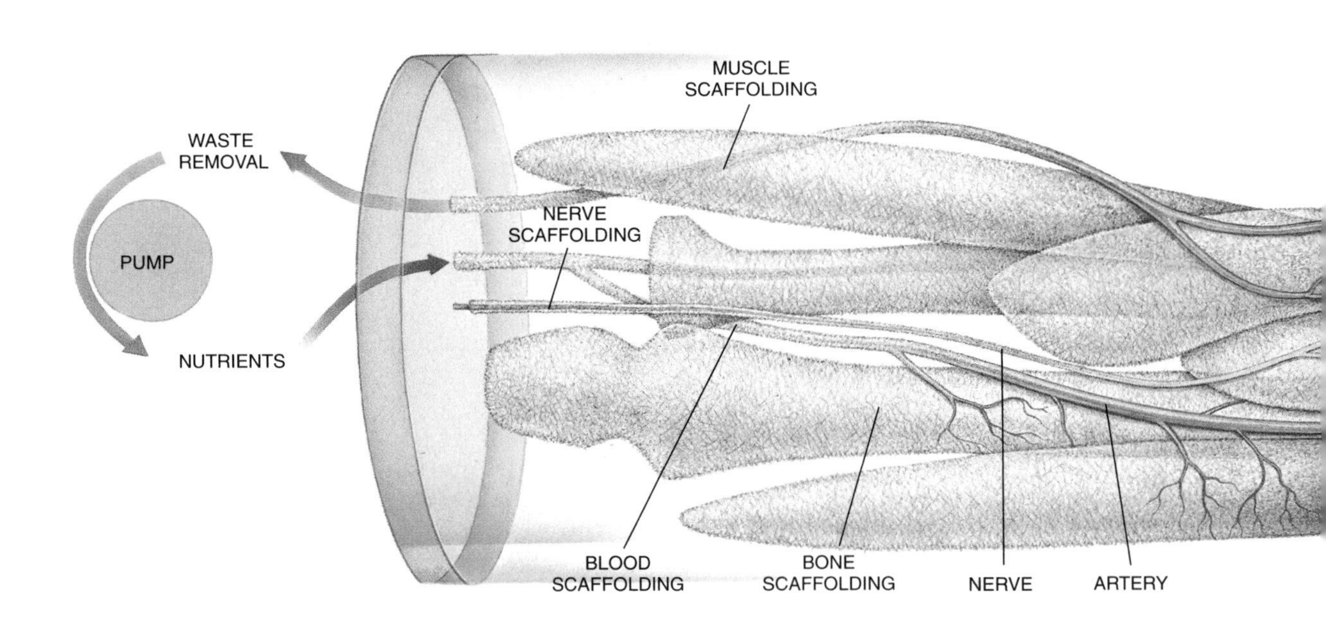

nerve cells. But a great deal of research is being devoted to this problem, and many investigators are confident that it will be overcome.

In the meantime, innovative microelectronic devices may substitute for implants of engineered nervous tissue. For example, a microchip implant may someday be able to restore some vision to people who have been blinded by diseases of the retina, the sensory membrane that lines the eye. In two of the more common retinal diseases, retinitis pigmentosa and macular degeneration, the light-receiving ganglion cells of the retina are destroyed, but the underlying nerves that transmit images from those cells to the brain remain intact and functional.

## Innovative Electronics

An ultrathin chip, placed surgically at the back of the eye, could work in conjunction with a miniature camera to stimulate the nerves that transmit images. The camera would fit on a pair of eyeglasses; a laser attached to the camera would both power the chip and send it visual information via an infrared beam. The microchip would then excite the retinal nerve endings much as healthy cells do, producing the sensation of sight. At M.I.T. and the Massachusetts Eye and Ear Infirmary, recent experiments in rabbits with a prototype of this "vision chip" have shown that such a device can stimulate the ganglion cells, which then send signals to the brain. Researchers will have to wait until the chip has been implanted in humans to know whether those signals approximate the experience of sight.

Mechanical devices will also continue to play a part in the design of artificial organs, as they have in this century. They will be critical components in, say, construction of the so-called artificial womb. In the past few decades, medical science has made considerable progress in the care of premature infants. Current life-support systems can sustain babies at 24 weeks of gestation; their nutritional needs are met through intravenous feeding, and ventilators help them to breathe.

Younger infants cannot survive, primarily because their immature lungs are unable to breathe air.

A sterile, fluid-filled artificial womb would improve survival rates for these newborns. The babies would breathe liquids called perfluorocarbons, which carry oxygen and carbon dioxide in high concentrations. Perfluorocarbons can be

**Figure 9.3 REPLACEMENT ARM AND HAND will someday be fabricated of artificial tissues. The structure of each system—muscle, bone, blood vessels, skin and so on—would be duplicated in biodegradable plastic. These "scaffolds" would then be seeded with cells of the relevant tissue. The cells divide, and the plastic degrades; finally, only coherent tissue remains. A mechanical pump would provide nutrients and remove wastes until the arm, which would take roughly six weeks to grow, can be attached to the body. (Depiction of a gradient of tissue formation is for illustrative purposes; all the tissue would in fact form at the same time.)**

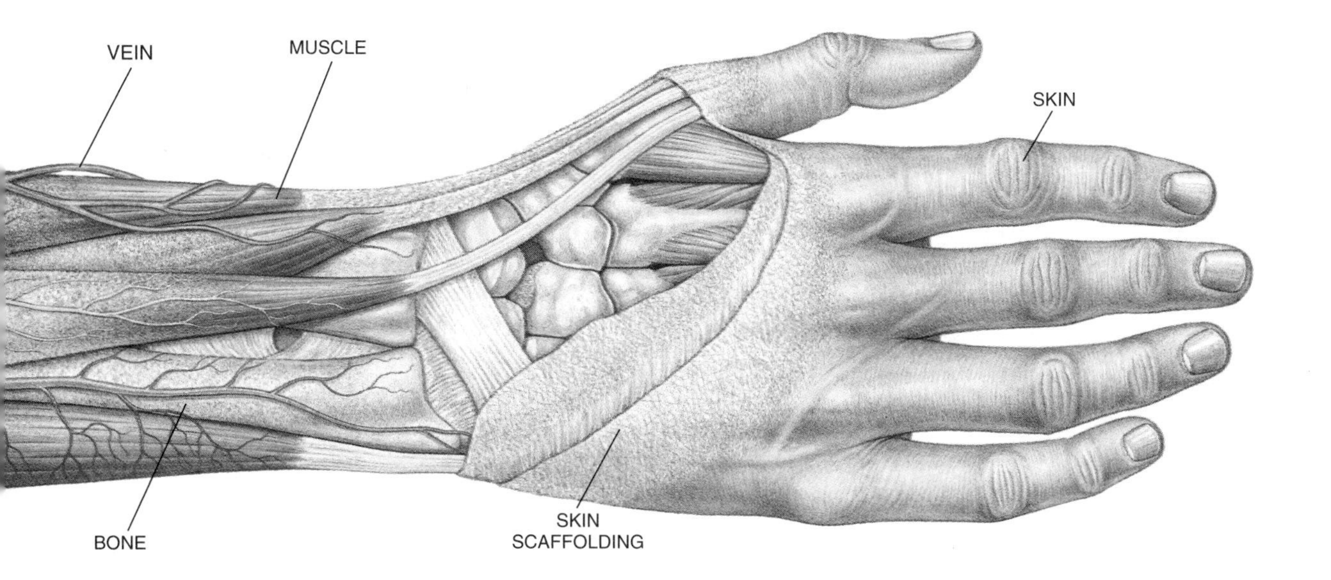

inhaled and exhaled just as air is. A pump would maintain continuous circulation of the fluid, allowing for gas exchange. Liquid breathing more closely resembles the uterine environment than do traditional ventilators and is much easier on the respiratory tract. Indeed, new work on using liquid ventilation in adults with injured lungs is under way. Liquid-ventilation systems for older babies are currently in clinical trials. Such systems will be used to sustain younger babies within a decade or so.

In addition to a gas-exchange apparatus, the womb would be equipped with filtering devices to remove toxins from the liquid. Nutrition would be delivered intravenously, as it is now. The womb would provide a self-contained system in which development and growth could proceed normally until the baby's second "birth." For most premature babies, such support would be enough to ensure survival. The developing child is, after all, the ultimate tissue engineer.

# Future Contraceptives

*Vaccines for men and women will eventually join new implants, better spermicides and stronger, thinner condoms.*

• • •

Nancy J. Alexander

With a range of contraceptives already available, are new ones even necessary? Absolutely. Among those who practice birth control (including more than half the world's couples), dissatisfaction with current methods is high. This unhappiness often leads to misuse or abandonment of otherwise effective approaches and thus contributes significantly to high rates of accidental pregnancy and of abortion. In the U.S., more than half of all pregnancies every year are unintended, and roughly 1.6 million of the approximately 6.4 million women who become pregnant elect to have an abortion.

The ideal contraceptive would be highly effective, safe, long-acting but readily reversible, virtually free of side effects and applied sometime other than just before sex. It would also reduce the spread of sexually transmitted diseases and be inexpensive. No product under study today can meet all these criteria, but several approaches that could be on the market in the first third of the 21st century would meet most of them.

## Beyond Condoms for Men

Right now the only contraceptive device sold for men is the condom. Once, men covered the penis with animal bladders and lengths of intestines. Later, the advent of latex allowed uniform products to be made, but many men claim the sheaths reduce their sexual pleasure. To address this powerful deterrent to use, manufacturers have recently introduced a strong, thin polyurethane condom, and additional films made of the same or other polymers will follow soon. Aside from interfering less with sensation, these nonlatex materials should be excellent barriers to infection, nonallergenic and more resistant to breakage and to degradation by heat, light and oily lubricants.

The first truly innovative male-oriented approach will manipulate hormones to stop sperm production—a big challenge, given that men typically generate at least 1,000 sperm every minute. Sperm manufacture is controlled by several hormones. The hypothalamus secretes gonadotropin-releasing hormone, which drives the pituitary to secrete luteinizing hormone and follicle-stimulating hormone. Luteinizing hormone stimulates the testes to produce testosterone. This steroid, together with follicle-stimulating hormone, induces cells called spermatogonia in the testes to divide and ultimately give rise to sperm.

One avenue of hormonal attack would be intramuscular injection of an androgen (testosterone or related male hormones), leading to release of the hormone into the bloodstream. This strategy, which is being studied intensively by the World Health Organization, derives from the finding that circulating androgens instruct the brain to dampen secretion of gonadotropin-releasing hormone and thus of luteinizing hormone and follicle-stimulating hormone. A drawback is that high levels of androgens in the circulation produce troubling side effects—notably, increased irritability, lowered levels of high-density lipoproteins (the "good" kind of cholesterol) and increased acne. Fortunately, adding a progestin (a synthetic form of the female steroid progesterone) seems to allow men to take a lower androgen dose, an innovation that should eliminate side effects and be safer than taking an androgen alone. The combination treatment will probably offer three months of protection and should be ready in about a decade. This lag time may seem surprising, but development of any new contraceptive technology takes 10 to 20 years.

As an alternative, one could block the activity of gonadotropin-releasing hormone with molecules that do not cause androgen-related side effects. Antagonists consisting of small proteins, or peptides, already exist, but they are not potent enough to serve as contraceptives. Investigators are therefore trying to design nonpeptide inhibitors. Of course, blockade of gonadotropin-releasing hormone would suppress testosterone production, and so men would have to take replacement androgens in order to retain muscle mass, male sexual characteristics and libido.

Perhaps 20 to 25 years from now, men should also have access to long-acting agents (protective for months) that, instead of disturbing natural hormone balances, directly halt sperm production in the testes or impede maturation of newly made sperm in matching chambers connected to the testes: the epididymides. It is in the epididymides, for example, that sperm first become motile. Investigations have suggested that disrupting maturation in the epididymides is the more feasible option, for a couple of reasons. Whether delivered orally, by injection or by an implant, drugs aimed at sperm would have to reach the testes or epididymides via the bloodstream; however, blood-borne drugs often have difficulty passing out of the circulation and into the part of the testes where sperm are manufactured. Further, many drugs that are capable of stalling sperm synthesis have proved toxic to spermatogonia in the testes and would thus lead to irreversible sterility. Such action would be fine for pets but would be unacceptable to most men.

## Vaccines for Both Sexes

The 21st century should also see contraceptive vaccines (immunocontraceptives) for men as well as women; these vaccines will probably be effective for about a year. Most of them would prod the immune system to make antibodies able to bind to, and disrupt the functioning of, selected proteins involved in reproduction. The immune system would be set in motion by injecting a person with many copies of the target protein—known as an antigen or immunogen—along with other substances capable of boosting the body's response.

One of the more promising vaccines to have reached clinical trials aims to raise antibodies that will inhibit gonadotropin-releasing hormone in men. Because this vaccine, developed by the Population Council, would shut down testosterone production, men would, once again, need androgen-replacement therapy.

**Figure 10.1 SPERM HEAD is the target of many contraceptive strategies under study for both sexes. Several approaches rely on synthetic chemicals (*red dots in detail of plasma membrane*) to alter the head in ways that would prevent sperm from fertilizing an egg. Scientists are also trying to produce contraceptive vaccines. Such vaccines would induce the immune system of a man or a woman to produce antibody molecules able to disrupt sperm function. Certain antibodies, for example, can cause sperm cells to clump together uselessly (*micrograph*).**

Two versions of a female vaccine are also being tested. They target human chorionic gonadotropin, a hormone that is secreted by the nascent placenta and is needed for implantation. One version, which was developed by the Indian National Institute of Immunology, has undergone preliminary human trials of efficacy in India, and similar trials are planned for the other approach, which is being developed by the World Health Organization.

Vaccines given to men or women could also lead to immune responses that would immobilize sperm, cause them to clump together or otherwise prevent them from swimming to or fusing with an egg (see Figure 10.1). Female vaccines could additionally stimulate the production of antibodies that would bind to the surface of an ovulated egg and form a shield impenetrable to sperm. Immunocontraceptives will be among the later birth-control

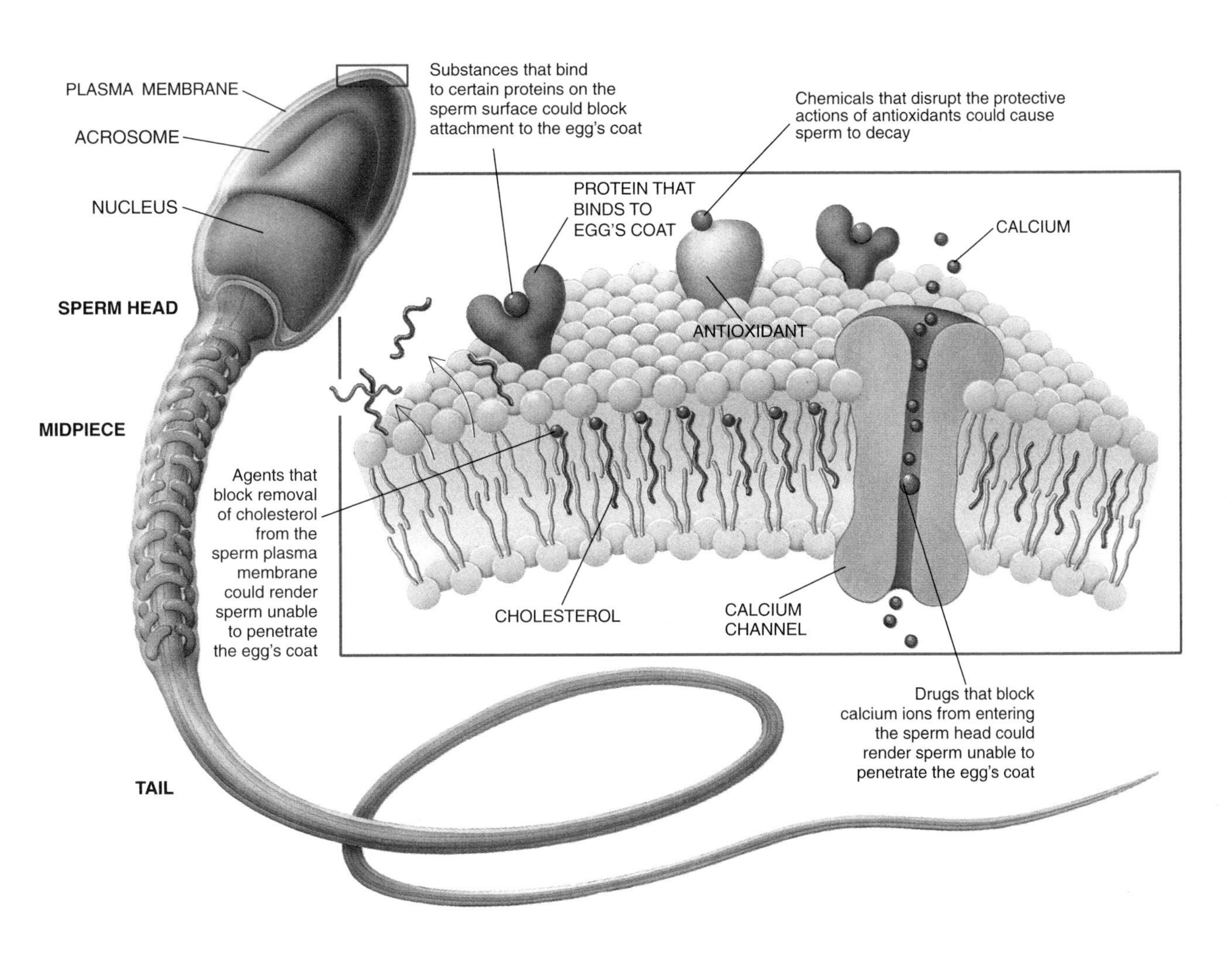

## Tomorrow's Infertility Treatments

Over the past decade investigators have markedly advanced techniques for in vitro fertilization—the procedure that yielded the first "test tube" baby in 1978. Today women can be induced to ovulate many eggs in a single menstrual cycle; the eggs can be removed and injected with sperm or DNA; the embryos can be screened for genetic defects; and healthy embryos can be implanted in the woman's womb. In the first third of the 21st century, researchers will probably refine these techniques and improve their success rates, but the basic approaches are unlikely to change radically.

In vitro fertilization is most suitable for women whose fallopian tubes are blocked or whose inability to conceive is unexplained. The many women who are infertile because of benign, fibroid tumors in the uterus need other therapies. At the moment, troublesome fibroids are usually reduced by surgery or administration of analogues of gonadotropin-releasing hormone. Surgery of any kind carries risks, and hormonal treatment can be problematic for infertile women, in part because when the analogues are in use they prevent conception; also, when treatment stops, the fibroids often return. If scientists can identify unique proteins on the tumors, they should be able to design drugs that will home in on those proteins and cure the tumors simply, without affecting other tissues or causing side effects. Chances are good that such drugs will become available by the year 2010.

Similar therapies will probably cure endometriosis, the growth of endometrial cells (from the lining of the uterus) outside of the uterus, such as in and around the fallopian tubes. The condition causes some infertility and can also lead to severe menstrual cramps.

But it is for men—particularly those who manufacture no sperm—that the most dramatic advances are likely to occur in the next 30 years. Scientists are perfecting ways to transplant donated precursors of sperm cells into the seminiferous tubules of the testes: the system of tubes in which sperm are ordinarily made. Normally, such precursors generate sperm throughout a man's lifetime; presumably the donated cells would do the same.

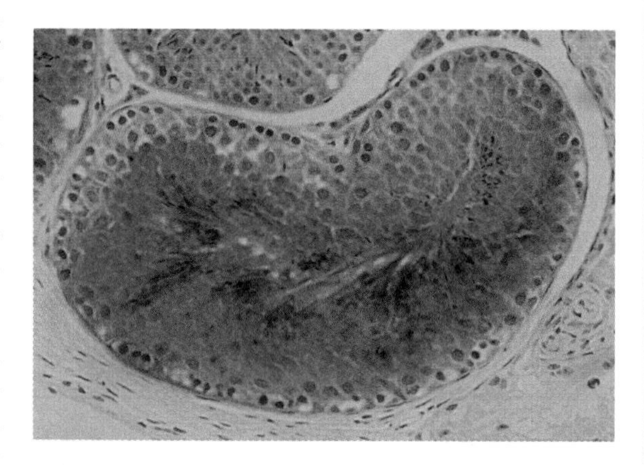

**MOUSE TESTIS that once made no sperm began manufacturing it in quantity after precursor cells from another mouse were implanted near the periphery. The wash of color in the center is a sign that the transplanted cells, which were stained blue, survived and gave rise to sperm.**

tools to appear, primarily because candidate immunogens must be studied in great detail. Scientists have to be sure that the inoculations will not induce immune responses against unintended tissues, and they have to find methods for producing vaccines in quantity. Adding to the complexity, vaccines will probably need to contain several antigens, to compensate for the fact that people can differ in their responsiveness to individual antigens.

### For Women: Many Hormonal Options

Well before vaccines appear and other alternatives to condoms show up for men, women will already have amassed experience with several new products, many of them hormonal. Currently women can choose from various hormonal methods: oral contraceptives, monthly injections, an injection that works for three months, and the five-year Norplant system, in which six flexible hormone-containing rods (the size of wooden matches) are placed under the skin. Birth-control pills, which are reasonably safe and very effective (and may well be sold without a prescription in the 21st century), provide progestins alone or in combination with synthetic estrogen. The injectables and implants consist of progestins. In all cases, the approaches serve to block ovulation and increase the thickness of cervical mucus so that sperm have difficulty navigating to the egg.

Within five years, women can expect to have a doughnut-shaped ring that will fit in the vagina like a diaphragm and release a progestin alone or with estrogen. A typical ring might be kept in place for three weeks and removed for one week (to allow for menses). It would thereby obviate the need to take a daily pill.

Also in the next few years, the manufacturer of Norplant is expected to introduce a second-generation implant consisting of just two rods that will be simpler than the original to insert and remove. In time there will be single-rod implants and a biodegradable system that will eventually dissolve in the body but will be removable, and hence reversible, for a while.

Soon after the year 2000 an intrauterine device (IUD) that releases a progestin and lasts for five

## When to Expect New Contraceptives

The dates below are estimates for the U.S.; the timing in other nations will depend in part on regulatory requirements.

| | By 2000 | By 2005 | After 2005 |
|---|---|---|---|
| FOR MEN | | | |
| | • Thinner, stronger nonlatex condoms | • Three-month injectable androgen with progestin | • Contraceptive vaccine that is effective for a year (by 2015)<br>• Injectable agents that block maturation of sperm and are effective for months (by 2015 or beyond) |
| FOR WOMEN | | | |
| | • Hormone-releasing implants that are easy to insert and remove and are effective for three to five years<br>• Hormone-releasing vaginal ring | • Progestin-releasing intrauterine device that works for five years (already sold in parts of Europe)<br>• New "morning-after" pills<br>• Spermicides that reduce the risk of acquiring sexually transmitted diseases | • Biodegradable implants that are effective for three to five years (by 2010)<br>• Contraceptive vaccines that are effective for a year (by 2015)<br>• Vaginal agents that block the union of sperm and egg (by 2015 or beyond)<br>• Possibly once-a-month pills (beyond 2015) |

years should be obtainable in the U.S. (It is already available in some European countries.) Existing nonhormonal IUDs work for many years and are quite effective in women who have been pregnant. But clinical tests indicate the new device produces less cramping and blood loss during menstruation and carries a lower risk of pelvic inflammatory disease.

Many researchers are giving special emphasis to developing a pill that could be taken just once a month, either at the time menses is expected or on the last day of a woman's period (to prevent pregnancy for the following month). For certain women, the latter would be most desirable, because it would prevent conception rather than induce an early abortion. Unfortunately, progress on such pills has been slowed by lack of detailed knowledge about the mechanisms that stimulate and stop monthly bleeding.

At some point in her life, a woman is likely to have unprotected sex. Emergency postcoital contraceptives are already sold; they are commonly known as morning-after pills, even though they can actually be taken two or three days after intercourse. Regular birth-control pills can also serve as emergency contraceptives if taken in the right doses. Yet all these approaches have unpleasant side effects, such as nausea, and so research is under way to develop gentler versions. For instance, the French pill RU 486, which is available in several nations outside the U.S. and is best known for terminating confirmed pregnancies, has been found to cause relatively little discomfort when taken in a low dose within 72 hours after unprotected sex.

## Sperm Police Could Act Locally

A rather different contraceptive strategy for women will rely on chemicals that directly prevent sperm from migrating to or fertilizing an egg. These compounds might be supplied by a vaginal ring, for long-term protection, or might be delivered into the vagina shortly before sex in the same way as spermicides are administered. They would differ from spermicides in several ways, however.

Spermicides are detergents that degrade sperm. Regrettably, these detergents can also kill microorganisms that help to maintain proper acidity in the vagina. And they can irritate the vaginal wall, thereby possibly facilitating infection by viruses and bacteria. The compounds under investigation would not be detergents; they would act more specifically, by interfering with selected events that follow ejaculation. Consequently, they could probably be formulated in ways that would be nonirritating and would avoid disrupting the vaginal flora.

The chemicals could impede any number of events that follow ejaculation. After sperm are deposited in the vagina, they undergo a variety of changes important for fertilization. For instance, cholesterol is removed from the membrane encasing the sperm head; this step makes the membrane more fluid and thus enables other molecules to move to new positions as needed. Later, with the assistance of progesterone in the fallopian tubes, specialized channels in the sperm membrane open, allowing calcium ions from the environment to flow in. These ions facilitate the "acrosome" reaction, which occurs when sperm first meet the zona pellucida, the jellylike coat surrounding the outer membrane of the egg. In the acrosome reaction the sperm plasma membrane fuses with that of the acrosome, a sac of enzymes in the sperm head. Then the enzymes pour out and carve a path for sperm through the egg's coat.

Chemicals being explored for impeding maturation and the acrosome reaction include ones that clog calcium channels or prevent cholesterol from leaving the plasma membrane. (These chemicals are also being eyed for men; they would be useful if they bound tightly to sperm in the epididymides and remained attached during the journey through the female reproductive tract.) Rather than preventing maturation of sperm, some researchers are attempting to induce a premature acrosome reaction—an event that would render sperm unable to meld with an egg.

Related strategies would inhibit specific interactions that normally occur between sperm and the zona pellucida. For instance, researchers at Duke University Medical Center have developed a compound that binds to a sperm protein (zona receptor kinase) at the site normally reserved for interaction with a protein in the zona. Attachment of the compound blocks the enzyme from acting on the egg's coat. Additionally, sperm and seminal fluid carry antioxidant enzymes, which protect the integrity of the sperm membrane. If certain of these antioxidants were unique to sperm or to seminal fluid, investigators would be able to design drugs that inactivated those particular enzymes but did not deprive other cells of antioxidant defenses. In some instances, pure, laboratory-generated antibodies (monoclonal antibodies) might take the place of certain nonbiological chemicals as the active ingredients in contraceptives delivered into the vagina.

## Combating Sexual Disease

As the incidence of sexually transmitted diseases climbs, it is becoming increasingly critical to combine contraception with barriers to infection. Such barriers can be physical (as in male and female condoms and, to some degree, the diaphragm) or chemical (as in spermicides or the more selective formulations under investigation). To prevent infection by the AIDS-causing human immunodeficiency virus or other microbes, a chemical barrier would, at minimum, have to coat the entire vaginal wall and cervix, be nonirritating and be nontoxic to beneficial microbes in the vagina. These features alone might be enough to reduce infection significantly, but agents intended to kill harmful microorganisms actively could be added as well.

Investigators are only now beginning to evaluate whether existing spermicides possess the essential features needed to avoid transmission of sexually transmitted diseases. So far it appears the formulations do reduce the risk of acquiring at least one sexually transmitted disease—*Chlamydia* infection. By the year 2005, spermicides designed to include disease-preventing properties will probably be sold. Such spermicides would be of enormous benefit to female health because they would offer unobtrusive protection when men fail to use condoms and when condoms themselves fail.

## A Time Line

Obviously, the contraceptive methods that are emerging today are the fruits of research that was initiated years ago. The only new products that will be attainable before the year 2000 are nonlatex condoms, vaginal rings and, possibly, implants for women consisting of one or two hormone-filled rods that are easier to remove than are the extant capsules. Within a few years thereafter, spermicides that reduce the spread of sexually transmitted diseases will be a welcome addition, as will the progestin-releasing intrauterine device.

The next contraceptives to arrive in the U.S. should be, in order of appearance, new emergency contraceptives (by 2005), a three-month injectable androgen plus a progestin for men (also by about 2005), biodegradable implants for women (by 2010) and immunocontraceptives (perhaps by 2015). (Introduction times outside the U.S. will be influenced by the regulatory requirements of each nation.)

Actually, immunocontraceptives and other technologies in early stages of development will reach the market only if the pharmaceutical and biotechnology industries take more interest in them. Most of them are being explored with seed money from the U.S. government and nonprofit organizations here and abroad. But those funds, totaling about $57 million annually (less than a fourth the cost of developing a single product), cannot pay for the extensive trials of safety and efficacy that are required for licensing. Such tests require the backing of private industry—support that may not become strong until limits on liability are set and a misperception that little money can be made from contraceptives is overcome.

Because no contraceptive is perfect for everyone, people who want to practice birth control need a broad array of choices. In spite of the obstacles, I am optimistic that today's ongoing work will go far toward meeting that need in the century ahead.

## Further Predictions on Medical Progress

*On April 25, 1995, Scientific American convened experts to speculate on the key medical technologies of the 21st century and the challenges arising from them. Some excerpts follow:*

**FRANCIS S. COLLINS, director of the National Center for Human Genome Research at the National Institutes of Health and one of the discoverers of the gene that causes cystic fibrosis:**

People are increasingly interested in taking more responsibility for their own medical care and for that of their loved ones. It is reasonably likely that by the year 2010, when you reach your 18th birthday you will be able to have your own report card printed out of your individual risks for future disease based on the genes you have inherited. I suspect many people will be interested in that information, particularly if it is focused on diseases where alterations in lifestyle and medical surveillance can reduce that risk to a more manageable level.

In addition to being able to predict risk for disease quite early, two other consequences of gene discovery will be rather important. One will be the ability to move early detection of actual disease to an earlier and earlier stage, particularly for cancer. At the moment, by the time you have a positive mammogram for breast cancer, you have had that tumor a long time; the chance that the cancer has already spread is substantial. If we could come up with molecular probes that highlighted the first few cells when they moved down the pathway toward cancer, then the probability of successfully treating that disease would become drastically better.

Also, there is growing confidence—well justified by research in both academia and the pharmaceutical industry—that the time is coming when there will be magic bullets to treat cancer the way we now treat many infectious diseases with vaccines and antibiotics. Our understanding of oncogenes and tumor suppressors and of the detailed steps that carry a cell from being normal to being malignant is going to allow us to develop drugs that will make our current chemotherapy poisons look just as obsolete as arsenic now is for treating infectious illnesses. And I think that time is coming potentially within the next 20 to 25 years.

**ROBERT E. McAFEE, then president of the American Medical Association:**

There is going to be a lot of "gadgeteering" in the next 25 years. I think you are going to see a bedside device that will do 100 different blood tests on a single drop of blood—a process that now normally takes hours, if not days, to do. In the vascular surgery field, new therapies will make some abused arteries whole again, at least for a time.

Unfortunately, at the end of 25 years, death still will not be optional. We appear to be the only country that looks upon death as an option, and I'm afraid that isn't going to be available to us.

**STEVEN A. ROSENBERG, chief of surgery at the National Cancer Institute, professor of surgery at the George Washington University School of Medicine and editor of *The Cancer Journal from Scientific American:***

My crystal ball tells me that an increased understanding of the immune system and an increased ability to manipulate it genetically will have a major impact in the next 10 to 20 years, in at least two ways. The body's immune system recognizes invaders as things to destroy. If we can learn to control such destruction, then the whole problem of defective organs can be eased profoundly by transplantation of animal tissues and organs—that is, by xenotransplanation. Xenotransplantation, an area in which substantial progress is being made as we sit here, may enable us to get new kidneys, new hearts, new livers. Today a major stumbling block to the development of transplantation as a way to cure increased numbers of individuals is the low availability of organs. If we can figure out ways to keep animal organs from being rejected, we may be able to obtain organs from experimental animals, probably mini-pigs that will be raised specifically as a source of organs to replace human tissues.

On the other side, of course, is figuring out ways to stimulate the immune system to eliminate foreign invaders and cancers that we cannot now destroy. During the past decade, we have had the first successful applications of immunotherapies for cancer. For certain cancers, we can actually manipulate the immune system so that the cancers can be rejected by the body rather than by the application of an external treatment. As we learn more about the immune system, this approach to cancer treatment probably is going to hold the key to major improvements as well.

**JOHN F. BEARY III, senior vice president for regulatory and scientific affairs at Pharmaceutical Research and Manufacturers of America:**

Technologies are emerging that should help to address problems associated with aging and degenerative disorders such as osteoarthritis. It is possible that we will be able to grow cartilage again and that knee and hip diseases, which greatly detract from the quality of life in the later years, can be dealt with. The chemistry of the bone will be much better understood, and, since osteoporosis in women is such a common problem, that knowledge will have a tremendous impact on successful aging.

**FLORENCE P. HASELTINE, director of the Center for Population Research at the National Institutes of Health:**

I think we will be able to maintain a tremendous amount of our mobility as long as we can literally get out of the house. Right now a wheelchair that gets you around—one of those mechanisms you see scooting around—costs some $20,000, which is more than most of our cars cost. But those chairs will come down in price and be much more user-friendly. We are going to see a lot of interaction between the medical community, the biotech community and the robotics industry to make something you might call an intelligent wheelchair—that is, one that will sense where everything is in the room, sense when we should be eating and what we should be eating, things like that.

**SUSAN J. BLUMENTHAL, deputy assistant secretary for women's health, assistant surgeon general and rear admiral of the U.S. Public Health Service, and clinical professor of psychiatry at the Georgetown University School of Medicine:**

If we were to walk through a 19th-century graveyard, we would find that on average Americans died at age 48. In this century we have extended the life span for Americans by 30 years. We used to die of acute illness, and among women, also from childbirth. Today we die of chronic illnesses (such as heart disease, cancer and diabetes) and also from accidents, homicide and suicide. In more than 50 percent of these leading causes of death, behavioral and lifestyle factors are paramount: poor diet, lack of exercise, failure to wear seat belts, guns, substance abuse and smoking. If I were to tell you that we had discovered a new medical intervention that could decrease premature deaths in this country by one half and could cut chronic disability by two thirds and acute disability by one third, everyone would be saying, What is that new technology? What is that new drug? The truth is, it is low-cost behavioral interventions that can decrease the risk of these illnesses.

I am optimistic that we will, with an investment in behavioral research, be able to develop more precise behavior-changing strategies for diverse groups of people. These strategies will hold tremendous promise for decreasing premature morbidity and mortality in the century to come.

For mammographies, we are still using a 40-year-old technology to find small lumps in women's breasts. Medical imaging is 10 years behind intelligence and defense imaging. Let's get a peace dividend from our national investment in defense by applying these technologies to medical purposes, such as improving early detection of diseases.

# Commentary: An Improved Future?

*Medical advances challenge thinking on living, dying and being human.*

• • •

Arthur Caplan

It is not hard to divine a future in which technology improves the quality of life. Advances in medicine promise healthier lives through therapy in the womb, genetic manipulation, construction of artificial organs, use of designer drugs and application of other ingenious techniques for restoring organ function. Nor is it difficult to imagine life spans being extended through various tricks directed at thwarting the assaults of infectious agents and at eliminating the ailments of modern life.

Despite these rosy prospects for increasing a person's chance of being born healthy and living longer, many in the ethics guild—which is inclined to hand-wringing—wear frowns today. Tomorrow's medicine may enable us to live longer, but, these ethicists ask, will it enable us to live better? They note, for instance, that prevailing demographic and fiscal trends foretell little but misery.

They contend that as technology saves and extends lives, the prospects for our descendants become nastier. Longer life spans will cause an already struggling planet to be populated with a bumper crop of doddering elders. As medicine becomes better at rescuing imperiled lives and extending mortally threatened ones, the strains on natural resources, the diversion of economic resources to health care needs and the likelihood of international as well as intergenerational conflict should all increase. Plain old human misery will also rise, as more and more of us make it to our seventies, eighties and nineties, only to greet the arthritis, stroke, Alzheimer's or Parkinson's disease that our technological arrogance has in store for us.

The grim scenarios extend further. Some experts argue that strict rationing of high technology must soon become an unfortunate fact of life in economically developed nations. (The residents of less developed countries exit life more quickly; consequently, their governments will be spared the harsh decisions about which citizens should be tossed from the collective lifeboat.) Others worry that if the Americans, Swedes, Germans, Canadians, Norwegians, Italians, Taiwanese and Japanese keep spending money on health care at the current rate, the entire gross national products of these nations soon will be devoted to nothing else. Still others are petrified at the prospect of human beings so greedy for more and more life that they are willing to rob the purses of their own children and grandchildren to pay for this supremely suspect indulgence.

Admittedly, these forecasts reveal some important dangers of rapid advances in medicine. Yet such concern may be unwarranted. Medical progress has thus far proved enormously expensive and morally ambivalent, extending life but often ignoring its quality. Yet the technology of the 21st century need not follow this path.

If, as seems likely, medicine moves to diagnosis and treatment of disease at its molecular level, away from the contemporary practice of treating overt symptoms and organ failures, the costs of medical progress might actually decrease. It is very resource intensive—and thus very expensive—to transplant a liver from a cadaver to a child whose own organ is failing because of a congenital disease. Pediatric liver transplants require skilled surgery, long hospitalizations, many transfusions of blood, perpetual suppression of the patient's immune system and extended counseling for the family and child. Identifying fetuses at risk for congenital liver disease during pregnancy and then repairing the disorder by gene therapy should prove considerably cheaper, particularly if such therapy could be given once, early in life.

Similarly, institutionalizing people who have chronic schizophrenia, severe depression or extreme substance abuse problems is very expensive. Treatment with designer drugs made to order for these problems, with microchip implants that release needed chemicals into the body or perhaps with virtual-reality therapy, in which patients find relief from their condition in a computer-generated fantasy world, may well prove far less costly.

Indeed, if the latest clinical procedures (such as noninvasive, highly detailed magnetic resonance imaging of internal organs) as well as new technologies just beginning to appear (including genetic diagnostic tests and vaccines for chicken pox and hepatitis) are any indication, medicine could be both increasingly economical and less hazardous to patients. A great deal of cleverness in public policy and a fair amount of luck will be needed, but we could find ourselves facing a future in which more of us live longer and in better health for the vast part of our lives without having to spend outrageous sums.

Thus, although it is tempting to forecast nothing but financial ruin and societal despair in the wake of medical progress, this tendency ought to be resisted. Still, there are serious concerns that go beyond the standard problems of demographics and money. For example, if we can cure mental illness by altering the chemistry of our brains, do we risk losing our sense of identity in the process? If disease is diagnosed and treated most effectively by tinkering with our molecules, will anyone be capable of, or paid for, talking with patients? Or will a quick dose of a drug become all we expect from health care providers? Will society come to think it is irresponsible to bear children without having a thorough genetic physical first? As we learn more about how genes regulate the body's functions, who will decide whether characteristics such as short stature, baldness, albinism, deafness, hyperactivity or aggressiveness are classified as diseases rather than merely differences? And on what basis will these distinctions be made?

The very reductionism to the molecular level that is fueling the medical revolution also poses the greatest moral challenge we face. We need to decide to what extent we want to design our descendants. In grappling with this issue, we will come face-to-face with questions about the malleability and perfectibility of our species; this encounter will make current arguments over the relative influence of genetics and upbringing seem trivial.

Will the degree of philosophical angst be such that we ought to rein in the speed at which medicine, and especially gene therapy, proceeds? Is the emerging revolution in medical technology one that will leave us so befuddled about who we are and why we are here that we will one day look back longingly for simpler times when children died of illnesses such as Tay-Sachs disease, sickle cell anemia and cystic fibrosis, and adults were ravaged by cancer, diabetes and heart disease?

Hardly. If being human means using intelligence to improve the quality of life, there is little basis for ethical ambivalence or doubt about eliminating genetic scourges, much as we have done with infectious diseases such as polio and smallpox. Who would argue that our freedom from those diseases that devastated past populations has made us feel less human? Changing our biological blueprints in pursuit of longer, healthier lives should ideally pose no more of a threat than does taking impurities out of the water supply to protect human health.

This logic also extends to the most awe-inspiring option that will ultimately be available: the genetic modification of reproductive, or germ, cells. Changes to these cells alter the DNA of those who receive the therapy as well as the DNA of their descendants. Germ-cell therapy permanently changes the human gene pool, but the benefits of forever eliminating diseases such as spina bifida, anencephaly, hemophilia and muscular dystrophy would seem to make germ-cell gene therapy a moral obligation. We have been altering our gene pool for millennia through wars, selective mating, better diet and the evolution of medicine. Changing the genes of our offspring directly through genetic engineering is really a difference in degree rather than in kind of treatment. Yet we must guard diligently against misuse of the ability to design our descendants with careful regulation, legislation and societal consensus.

History is a funny thing. There is always a tendency to be nostalgic about times gone by. But the benefits of a longer life burdened by less disability, dysfunction and impairment will make the future as attractive as the past. If we can make that future affordable, if we can enhance the quality of our lives while controlling population growth, and if we can accept the idea that it is neither arrogant nor foolhardy to modify our biological constitution in the hope of preventing disorder and disease, medical technology could very well make the future better.

# PART IV
# MACHINES, MATERIALS AND MANUFACTURING

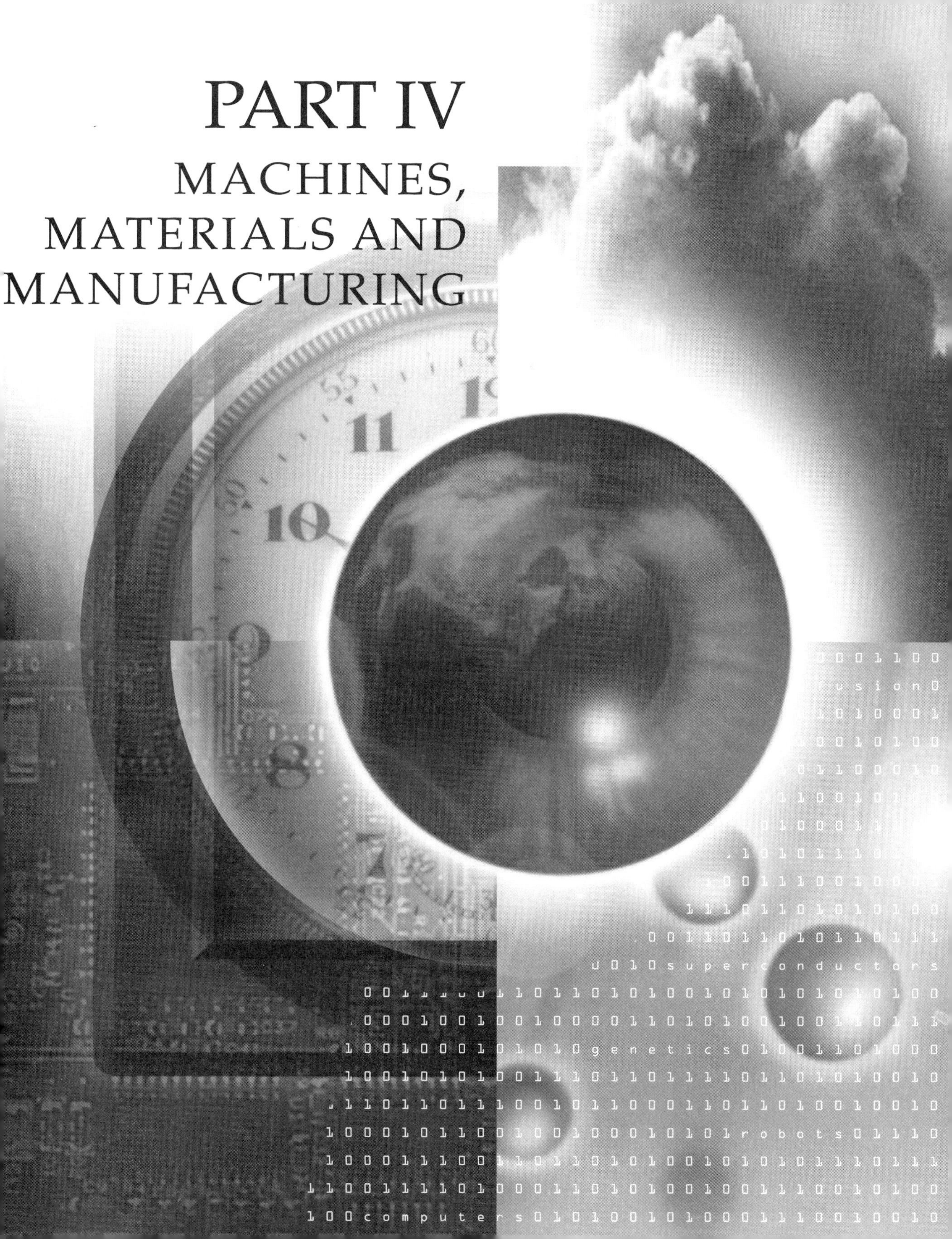

# Self-Assembling Materials

*The smaller, more complex machines*
*of the future cannot be built with current methods:*
*they must almost make themselves.*

• • •

George M. Whitesides

Our world is populated with machines, nonliving entities assembled by human beings from components that humankind has made. Our automobiles, computers, telephones, toaster ovens and screwdrivers far outnumber *us*. Despite this proliferation, no machine can reproduce itself without human agency. Yet.

In the 21st century, scientists will introduce a manufacturing strategy based on machines and materials that virtually make themselves. Called self-assembly, it is easiest to define by what it is not. A self-assembling process is one in which humans are *not* actively involved, in which atoms, molecules, aggregates of molecules and components arrange themselves into ordered, functioning entities without human intervention. In contrast, most current methods of manufacturing involve a considerable degree of human direction. We, or machines that we pilot, control many important elements of fabrication and assembly. Self-assembly omits the human hand from the building. People may design the process, and they may launch it, but once under way it proceeds according to its own internal plan, either toward an energetically stable form or toward some system whose form and function are encoded in its parts.

The concept of self-assembly is not new. It was inspired by nature, where entities as simple as a raindrop or as complex as a living cell arise from physical principles or instructions implicit in their components (see box "Two Types of Self-Assembly"). And it is already exploited in the manufacture of a number of common products. Most window glass, for example, is so-called float glass, made by floating molten glass on a pool of molten metal. The metal tends to minimize its surface area by becoming smooth and flat; consequently, the glass on top becomes optically smooth and uniformly flat as well. It is much less expensive to produce float glass than it is to grind and polish glass produced by other processes, and the quality of the float-glass surface is higher. Similarly, conventional manufacturing methods cannot specify the placement of the silicon and dopant atoms in a semiconductor crystal. The growth of the crystal from a melt of silicon is dictated by thermodynamic principles, not human presumption.

Such examples illustrate the potential of self-assembly. Those materials were arrived at almost by accident. In the next few decades, however, materials scientists will begin deliberately to design machines and manufacturing systems explicitly incorporating the principles of self-assembly. The approach could have many advantages. It would allow the fabrication of materials with novel properties. It would eliminate the error and expense introduced by human labor. And the minute machines of the future envisioned by enthusiasts of so-called nanotechnology would almost certainly need to be constructed by self-assembly methods.

## From Materials to Machines

The rational design of self-assembling machines begins with the rational design of self-assembling materials. The spherical, microscopic capsules known as liposomes are among the earliest successes (see Figure 11.1). Since the 1960s, biomedical researchers have been experimenting with liposomes as a vehicle for transporting drugs in the body; because the capsules protect their cargo from degradation by enzymes, a drug contained in a liposome envelope can remain active for longer periods than it would otherwise.

Liposomes were modeled after cell membranes, which are among the most striking examples of self-assembly in nature. Cell membranes are made mostly of molecules called phospholipids that have a kind of dual personality: one end of the phospholipid is attracted to water, and the other end is repelled by it. When placed in an aqueous environment, the molecules spontaneously form a double layer, or bilayer, in which the hydrophilic ends are in contact with the water and the hydrophobic ends point toward one another. Researchers use these same phospholipids to make liposomes. If there are enough molecules, the phospholipid bilayer will grow into a sphere with a cavity large enough to hold drug molecules. The liposomes are then injected into the body, and the drugs are released either by leakage or when a sphere ruptures. Liposome drug-delivery systems are currently in clinical trials.

## SAMs

The example of nature has helped to illuminate liposome research, but many investigations of self-assembling behavior must start almost from scratch. A self-assembled monolayer—affectionately referred to as SAMs by those who work with them—is a simple prototype that exemplifies the design principles materials researchers are exploring. A SAM is a one- to two-nanometer-thick film of organic molecules that form a two-dimensional crystal on an

**Figure 11.1 THE LIPOSOME (*left*) represents an early success among self-assembling materials: the microscopic capsules are in clinical trials as a drug-delivery system within the body. Current research focuses on self-assembling layers built from sausage-shaped molecules. At one end of each molecule is an atom that interacts strongly with a surface; at the other end, a variety of other atomic groupings. These molecules can organize themselves on one surface, thereby creating at their other end a surface with different properties. Such monolayers are being used to guide the growth of living cells. Additional layers can be built onto the monolayer; these self-assembling multilayers are being explored as coatings to control reflection in devices that use light in communications.**

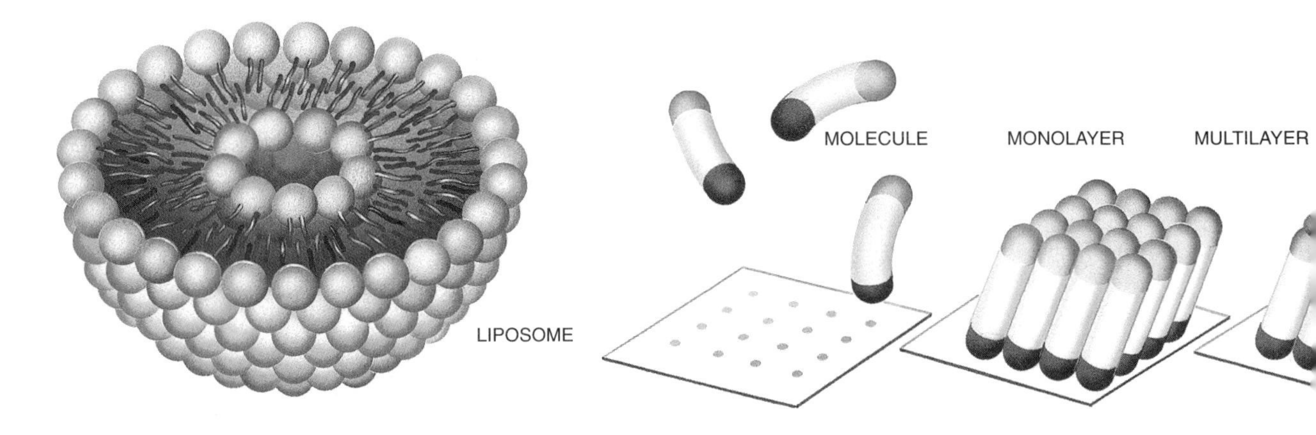

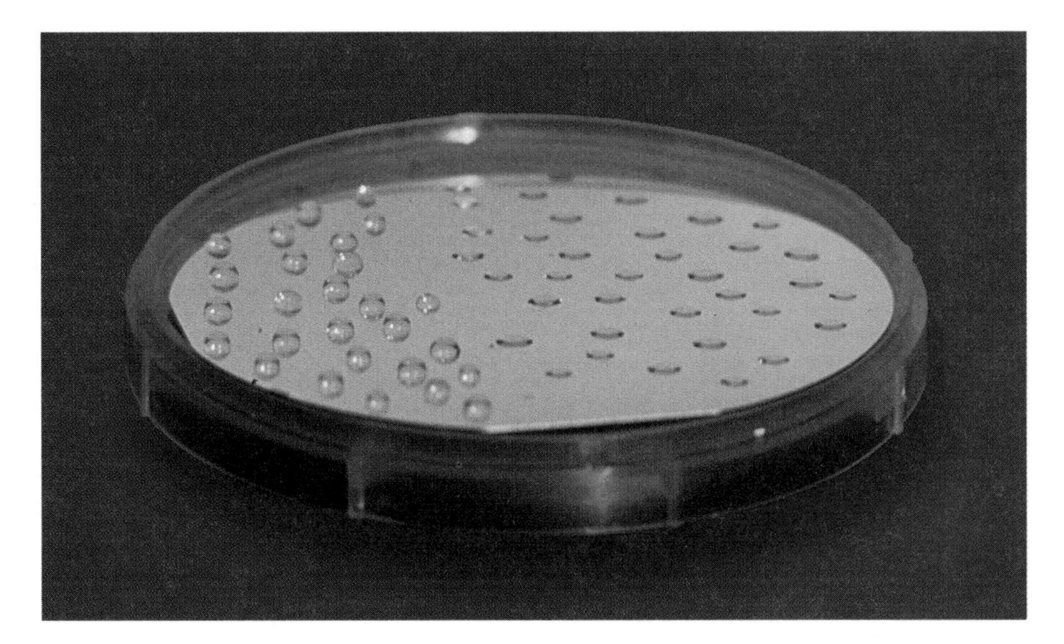

Figure 11.2 GOLD-COATED SILICON WAFER is often used in studying self-assembled monolayers. In this experiment the left half was covered with a monolayer having a hydrophobic surface, the right half with one presenting a hydrophilic surface. Drops of water flattened on the hydrophilic side but formed round beads that minimized contact with the surface on the hydrophobic side. The behavior shows that the outermost part of the self-assembled monolayer controls the wettability of the surface. The same strategy can be used to control adhesion, friction and corrosion.

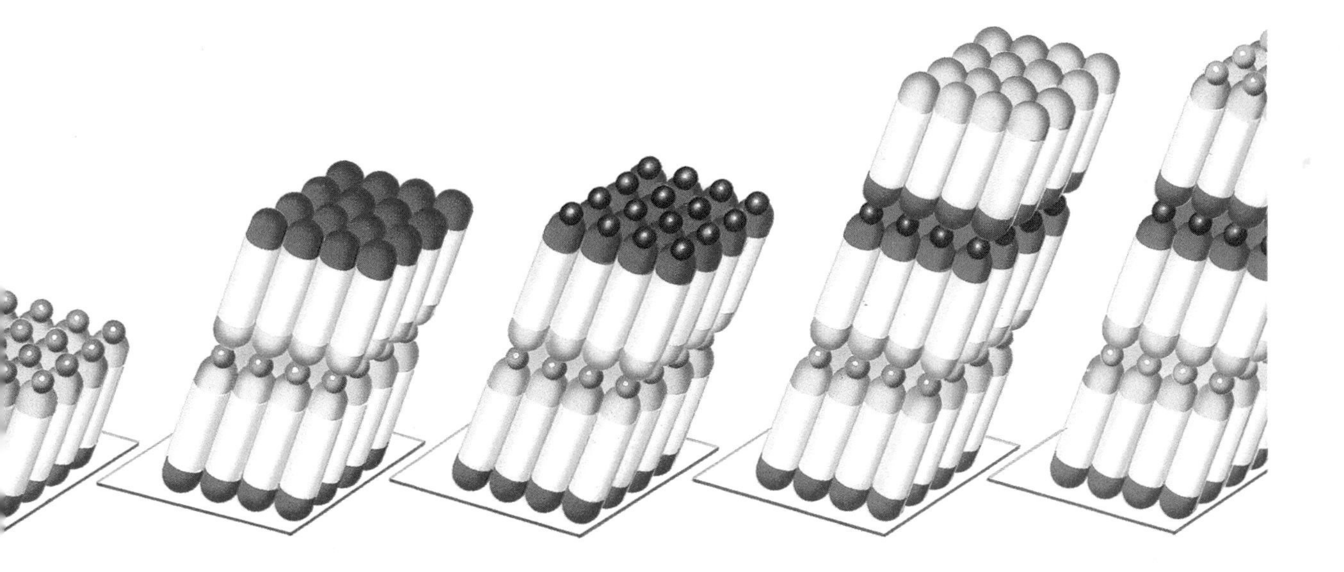

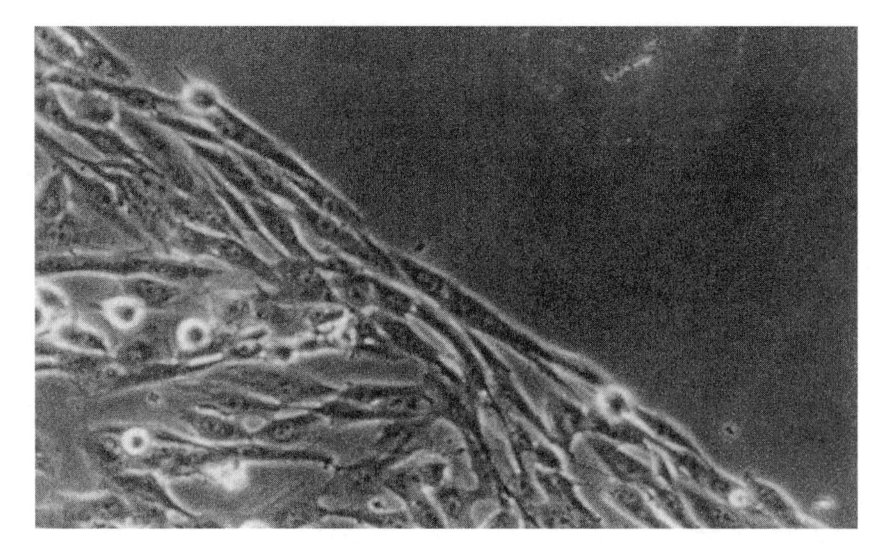

**Figure 11.3 CELL PATTERNING reveals the boundary between two different types of self-assembled monolayers. One side was designed to promote, the other to prevent, the adhesion of mammalian cells. The ability to control the attachment of living cells is being used to understand how cells interact with man-made surfaces and with surfaces that mimic those in living organisms and to improve the performance of devices that will be implanted in the body.**

adsorbing substrate. The molecules in a SAM are sausage-shaped, being longer than they are wide. At one end is an atom or group of atoms that interacts strongly with the surface; at the other, chemists can affix a variety of atomic groupings, thereby altering the properties of the new surface formed by the SAM.

The most extensively studied system of SAMs is made of molecules called alkanethiols, long hydrocarbon chains with a sulfur atom at one terminus. The sulfur adsorbs well onto a substrate of gold or silver. When, say, a glass plate coated with a thin film of gold is dipped into a solution of alkanethiol, the sulfur atoms attach to the gold. The distance between the sulfur atoms adsorbed on the surface is about the same as the cross-sectional diameter of the rest of the molecule (hence the sausage shape), and the alkanethiols pack together, generating what is essentially a two-dimensional crystal.

The thickness of this crystal can be controlled by varying the length of the hydrocarbon chain, and the properties of the crystal's surface can be modified with great precision. By attaching different terminal groups, for example, the surface can be made to attract or repel water, which in turn can affect its adhesion, corrosion and lubrication (see Figure 11.2). If the alkanethiols are stamped onto the gold in a particular pattern, they can be used to investigate the growth of cells (see Figure 11.3) on different organic substrates or to construct diffraction gratings for optical instruments. In contrast to most procedures for surface modification, all these operations are simple and inexpensive, requiring neither high-vacuum equipment nor lithography.

## Buckytubes

Self-assembly has also produced tiny graphite tubes that are among the smallest electrical "wires" ever made. These tubes are called buckytubes (see Figure 11.4), because they are structurally similar to the carbon buckyballs named for their resemblance to the geodesic domes of Buckminster Fuller. Buckytubes consist of several nested, concentric cylinders with nanometer-scale diameters, and because they are made of graphite—the most thermodynamically stable form of carbon at atmospheric pressure—they tend to form under conditions that allow carbon to move toward thermodynamic equilibrium.

In one process, a small drop of liquid metal is exposed at high temperature to a carbon source, such as benzene. The carbon dissolves on one face of the metal droplet and, for reasons no one quite understands, precipitates on the other. As it precipitates, it forms a circular tube of graphite whose diameter is fixed by the size of the metal drop. The tube grows from the drop continuously as carbon is fed to it. Because supports holding millions of metal drops are simple to prepare, millions of buckytubes can be grown simultaneously in a single reactor. The tubes are good conductors of electricity; although it is still not clear how to assemble them into coherent structures, chemists hope to use the tubes as dopants to increase electrical conductivity in polymers and other nonconducting materials and someday to build circuits with them.

## Two Types of Self-Assembly

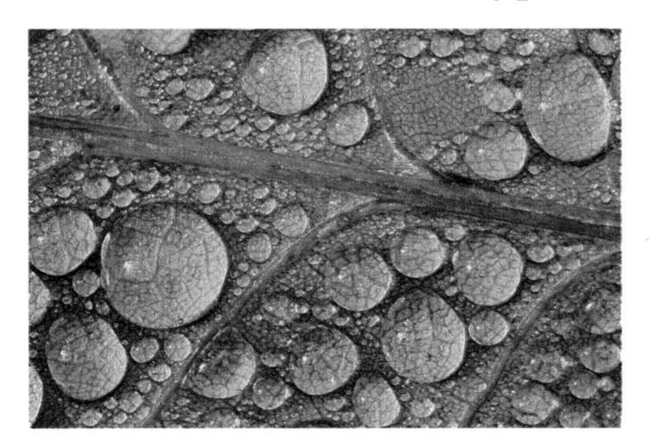

**RAINDROPS on a leaf illustrate thermodynamic self-assembly.**

Nature abounds with examples of self-assembly. Consider a raindrop on a leaf. The liquid drop has a smooth, curved surface of just the kind required for optical lenses. Grinding a lens of that shape would be a major undertaking. Yet the liquid assumes this shape spontaneously, because molecules at the interface between liquid and air are less stable than those in the interior. The laws of thermodynamics require that a raindrop take the form that maximizes its energetic stability. The smooth, curved shape does so by minimizing the area of the unstable surface.

This type of self-assembly, known as thermodynamic self-assembly, works to construct only the simplest structures. Living organisms, on the other hand, represent the extreme in complexity. They, too, are self-assembling: cells reproduce themselves each time they divide. Complex molecules inside a cell direct its function. Complex subcomponents help to sustain cells. The construction of a cell's complexity is balanced thermodynamically by energy-dissipating structures within the cell and requires complex molecules such as ATP. An embryo, and eventually new life, can arise from the union of two cells, whether or not human beings attend to the development.

The kind of self-assembly embodied by life is called coded self-assembly because instructions for the design of the system are built into its components. The idea of designing materials with a built-in set of instructions that will enable them to mimic the complexity of life is immensely attractive. Researchers are only beginning to understand the kinds of structures and tasks that could exploit this approach. Coded self-assembly is truly a concept for the next century.

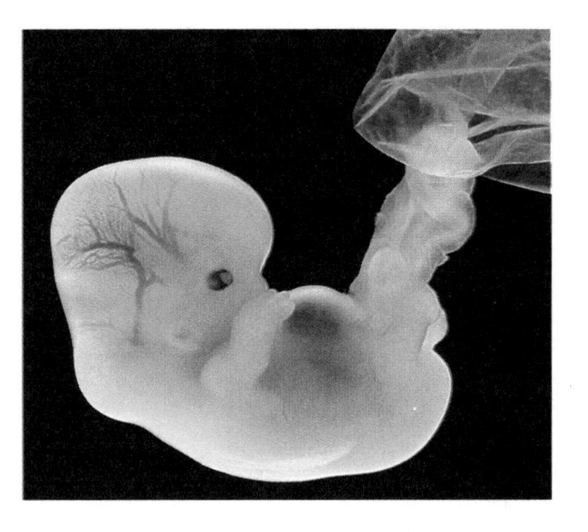

**EMBRYO exemplifies coded self-assembly.**

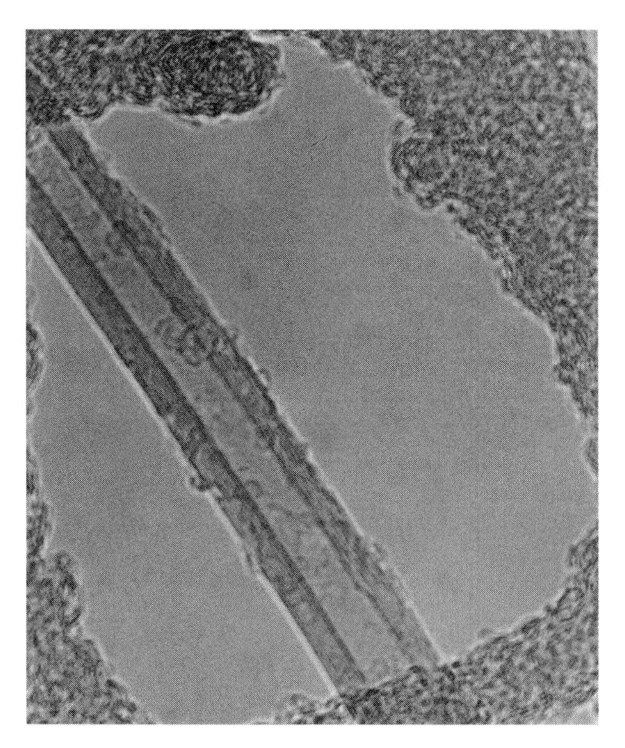

**Figure 11.4 BUCKYTUBES are among the smallest "wires" ever made. The electron micrograph shows in cross section the multiple layers of graphite of which they are composed. These nanometer-scale tubes are good conductors of electricity and may find use as dopants that increase conductivity in polymers and improve the performance of batteries.**

Indeed, buckytubes might be part of a more ambitious approach to self-assembly embodied in what materials researchers call a crystal memory: a self-assembling, three-dimensional version of the planar memories that are used in microelectronic devices today. At present, the crystal memory is purely conceptual: not one of its components has been demonstrated in the laboratory, even in principle. But they can be imagined.

The smallest unit of a crystal memory might be a silicon chip or some other semiconducting material that would be capable of carrying out a variety of microelectronic operations and that would have embedded in it instructions on how to behave when activated by signals from similar chips. These units would aggregate spontaneously, or crystallize, into a single larger unit, much as a liposome or SAM forms from smaller components. In their new configuration, the chips would stimulate one another and form electrical connections; signals from those connections would then trigger the units to begin differentiation, according to their embedded programming, into specialized roles: input or output units, switches, memory cells and so on.

If such a device seems improbable, consider that the process just described has many precedents in nature; in fact, all forms of life issue from simple subunits communicating among themselves. A microelectronic memory device could someday be able to build itself by the crystallization of smaller parts, thus ushering in a new era in manufacturing.

# Engineering Microscopic Machines

*Electronic fabrication processes can produce a data storage device or a chemical factory on a microchip.*

• • •

Kaigham J. Gabriel

The electronics industry relies on its ability to double the number of transistors on a microchip every 18 months, a trend that drives the dramatic revolution in electronics. Manufacturing millions of microscopic elements in an area no larger than a postage stamp has now begun to inspire technology that reaches beyond the field that produced the pocket telephone and the personal computer.

Using the materials and processes of microelectronics, researchers have fashioned microscopic beams, pits, gears, membranes and even motors that can be deployed to move atoms or to open and close valves that pump microliters of liquid. The size of these mechanical elements is measured in microns—a fraction of the width of a human hair. And like transistors, millions of them can be fabricated at one time.

In the next 50 years, this structural engineering of silicon may have as profound an impact on society as did the miniaturization of electronics in preceding decades. Electronic computing and memory circuits, as powerful as they are, do nothing more than switch electrons and route them on their way over tiny wires. Micromechanical devices will supply electronic systems with a much needed window to the physical world, allowing them to sense and control motion, light, sound, heat and other physical forces.

The coupling of mechanical and electronic systems will produce dramatic technical advances across diverse scientific and engineering disciplines. Thousands of beams with cross sections of less than a micron will move tiny electrical scanning heads that will read and write enough data to store a small library of information on an area the size of a microchip. Arrays of valves will release drug dosages into the bloodstream at precisely timed intervals. Inertial guidance systems on a chip will aid in locating the position of military combatants and direct munitions precisely to targets.

Microelectromechanical systems, or MEMS, is the name given to the practice of making and combining miniaturized mechanical and electronic components (see Figure 12.1). MEMS devices are made using manufacturing processes that are similar, and in some cases identical, to those required for crafting electronic components.

## Surface Micromachining

One technique, called surface micromachining, parallels electronics fabrication so closely that it is essentially a series of steps added to the making of a microchip. Surface micromachining acquired its name because the small mechanical structures are "machined" onto the surface of a silicon disk, known as a wafer. The technique relies on photolithography as well as other staples of the electronic manufacturing process that deposit or etch away small amounts of material on the chip.

Photolithography creates a pattern on the surface of a wafer, marking off an area that is subsequently etched away to build up micromechanical structures such as a motor or a freestanding beam. Manufacturers start by patterning and etching a hole in a layer of silicon dioxide deposited on the wafer. A gaseous vapor reaction then deposits a layer of polycrystalline silicon, which coats both the hole and the remaining silicon dioxide material. The silicon deposited into the hole becomes the base of the beam, and the same material that overlays the silicon dioxide forms the suspended part of the beam structure. In the final step, the remaining silicon dioxide is etched away, leaving the polycrystalline silicon beam free and suspended above the surface of the wafer.

Such miniaturized structures exhibit useful mechanical properties. When stimulated with an electrical voltage, a beam with a small mass will vibrate more rapidly than a heavier device, making it a more sensitive detector of motion, pressure or even chemical properties. For instance, a beam could adsorb a certain chemical (adsorption occurs when thin layers of a molecule adhere to a surface). As more of the chemical is adsorbed, the weight of the beam changes, altering the frequency at which it would vibrate when electrically excited. This chemical sensor could therefore operate by detecting such changes in vibrational frequency. Another type of sensor that employs beams manufactured with surface micromachining functions on a slightly different principle. It changes the position of suspended parallel beams that make up an electrical capacitor—and thus alters the amount of stored electrical charge—when an automobile goes through the rapid deceleration of a crash. Analog Devices, a Massachusetts-based semiconductor concern, manufactures this acceleration sensor to trigger the release of an air bag. The company has sold more than half a million of these sensors to automobile makers over the past two years (see Figure 12.2).

This air-bag sensor may one day be looked back on as the microelectromechanical equivalent of the early integrated electronics chips. The fabrication of beams and other elements of the motion sensor on the surface of a silicon microchip has made it possible to produce this device on a standard integrated-circuit fabrication line.

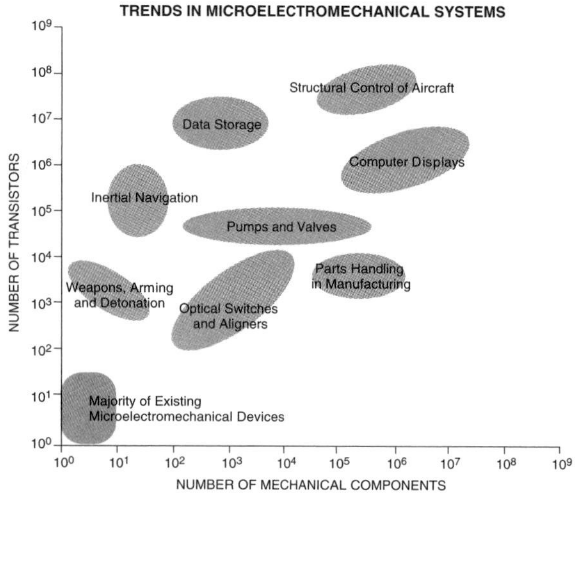

**Figure 12.1 EVOLUTION OF SMALL MACHINES AND SENSORS demonstrates that integrating more of these devices with electronic circuits will yield a window to the world of motion, sound, heat and other physical forces. The vertical axis shows information-processing ability; the horizontal axis indicates the devices' ability to sense and control. The green area represents devices that have already been developed; orange areas highlight future applications for the technology.**

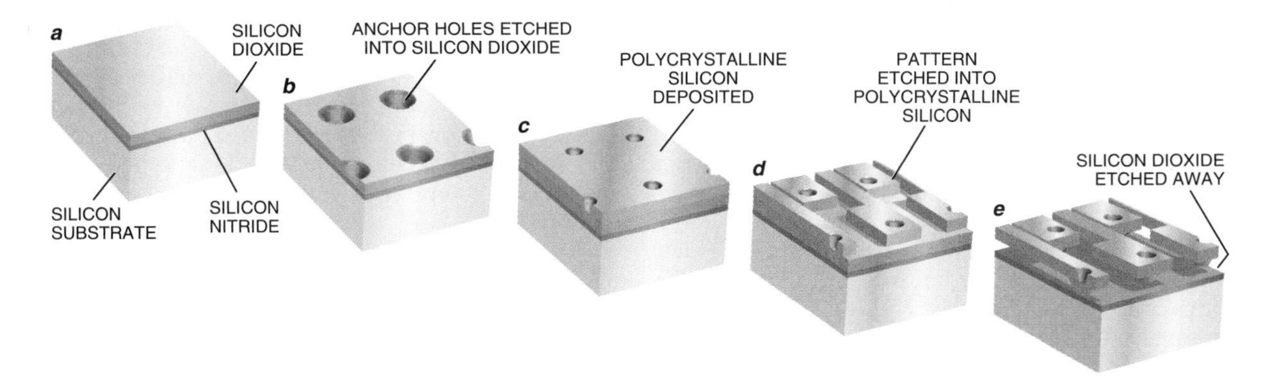


**Figure 12.2 BEAMS that serve as part of an acceleration sensor for triggering air bags (*photo*) are made by first depositing layers of silicon nitride (an insulating material) and silicon dioxide on the surface of a silicon substrate (*a*). Holes are lithographically patterned and etched into the silicon dioxide to form anchor points for the beams (*b*). A layer of polycrystalline silicon is then deposited (*c*). Lithography and etching form the pattern of the beams (*d*). Finally, the silicon dioxide is etched away to leave the freestanding beams (*e*).**

In microelectronics the ability to augment continually the number of transistors that can be wired together has produced truly revolutionary developments: the microprocessors and memory chips that made possible small, affordable computing devices such as the personal computer. Similarly, the worth of MEMS may become apparent only when thousands or millions of mechanical structures are manufactured and integrated with electronic elements.

The first examples of mass production of microelectromechanical devices have begun to appear—and many others are being contemplated in research laboratories all over the world. An early prototype demonstrates how MEMS may affect the way millions of people spend their leisure time in front of the television set. Texas Instruments has built an electronic display in which the picture elements, or pixels, that make up the image are controlled by microelectromechanical structures. Each pixel consists of a 16-micron-wide aluminum mirror that can reflect pulses of colored light onto a screen. The pixels are turned off or on when an electric field causes the mirrors to tilt 10 degrees to one side or the other. In one direction, a light beam is reflected onto the screen to illuminate the pixel. In the other, it scatters away from the screen, and the pixel remains dark.

This micromirror display could project the images required for a large-screen television with a high degree of brightness and resolution of picture detail. The mirrors could compensate for the inadequacies encountered with other technologies. Display designers, for instance, have run into difficulty in making liquid-crystal screens large enough for a wall-size television display.

The future of MEMS can be glimpsed by examining projects that have been funded during the past three years under a program sponsored by the U.S. Department of Defense's Advanced Research Projects Agency. This research is directed toward building a number of prototype microelectromechnical devices and systems that could transform not only weapons systems but also consumer products.

A team of engineers at the University of California at Los Angeles and the California Institute of Technology wants to show how MEMS may eventually influence aerodynamic design. The group has outlined its ideas for technology that might replace the relatively large moving surfaces of a wing—the flaps, slats and ailerons—that control both turning and ascent and descent. It plans to line the surface of a wing with thousands of 150-micron-long plates that, in their resting position, remain flat on the wing surface. When an electrical voltage is applied, the plates rise from the surface at up to a 90-degree angle. Thus activated, they can control the vortices of air that form across selected areas of the wing. Sensors can monitor the currents of air rushing over the wing and send a signal to adjust the position of the plates (see Chapter 6, "Evolution of the Commercial Airliner," by Eugene E. Covert).

These movable plates, or actuators, function similarly to a microscopic version of the huge flaps on conventional aircraft. Fine-tuning the control of the wing surfaces would enable an airplane to turn more quickly, stabilize against turbulence, or burn less fuel because of greater flying efficiency. The additional aerodynamic control achieved with this "smart skin" could lead to radically new aircraft designs that move beyond the cylinder-with-wings appearance that has prevailed for 70 years. Aerospace engineers might dispense entirely with flaps, rudders and even the wing surface, called a vertical stabilizer. The aircraft would become a kind of "flying wing," similar to the U.S. Air Force's Stealth bomber. An aircraft without a vertical stabilizer would exhibit greater maneuverability—a boon for fighter aircraft and perhaps also one day for high-speed commercial airliners that must be capable of changing direction quickly to avoid collisions.

## Miniature Microscopes

The engineering of small machines and sensors allows new uses for old ideas. For a decade, scientists have routinely worked with the scanning probe microscopes that can manipulate and form images with individual atoms. The most well known of these devices is the scanning tunneling microscope, or STM.

The STM, an invention for which Gerd Binnig and Heinrich Rohrer of IBM won the Nobel Prize in Physics in 1986, caught the attention of micromechanical specialists in the early 1980s. The fascination of the engineering community stems from calculations of how much information could be stored if STMs were used to read and write digital data. A trillion bits of information—equal to the text of 500 *Encyclopædia Britannica*'s—might fit into a square centimeter on a chip by deploying an assembly of multiple STMs.

The STM is a needle-shaped probe, the tip of which consists of a single atom (see Figure 12.3). A current that "tunnels" from the tip to a nearby conductive surface can move small groups of atoms, either to create holes or to pile up tiny mounds on the silicon chip. Holes and mounds correspond to the zeros and ones required to store digital data. A sensor, perhaps one constructed from a different type of scanning probe microscope, would "read" the data by detecting whether a nanometer-size plot of silicon represents a zero or one.

Only beams and motors a few microns in size, and with a commensurately small mass, will be

**Figure 12.3 MICROMECHANICAL INSTRUMENT consists of a motor and tip. The tip is visible on the suspended structure in the center of the motor; a similar tip is shown in close-up at the upper right. One day such motors, which measure 200 microns on a side (no more than two hair breadths), may maneuver tips to read and write data.**

able to move an STM quickly and precisely enough to make terabit (trillion-bit) data storage on a chip practicable. With MEMS, thousands of STMs could be suspended from movable beams built on the surface of a chip, each one reading or writing data in an area of a few square microns. The storage medium, moreover, could remain stationary, which would eliminate the need for today's spinning-media disk drives.

Noel C. MacDonald, an electrical engineering professor at Cornell University, has taken a step toward fulfilling the vision of the pocket research library. He has built an STM-equipped microbeam that can be moved in either the vertical and horizontal axes or even at an oblique angle. The beam hangs on a suspended frame attached to four motors, each of which measures only 200 microns (two hair widths) across. These engines push or

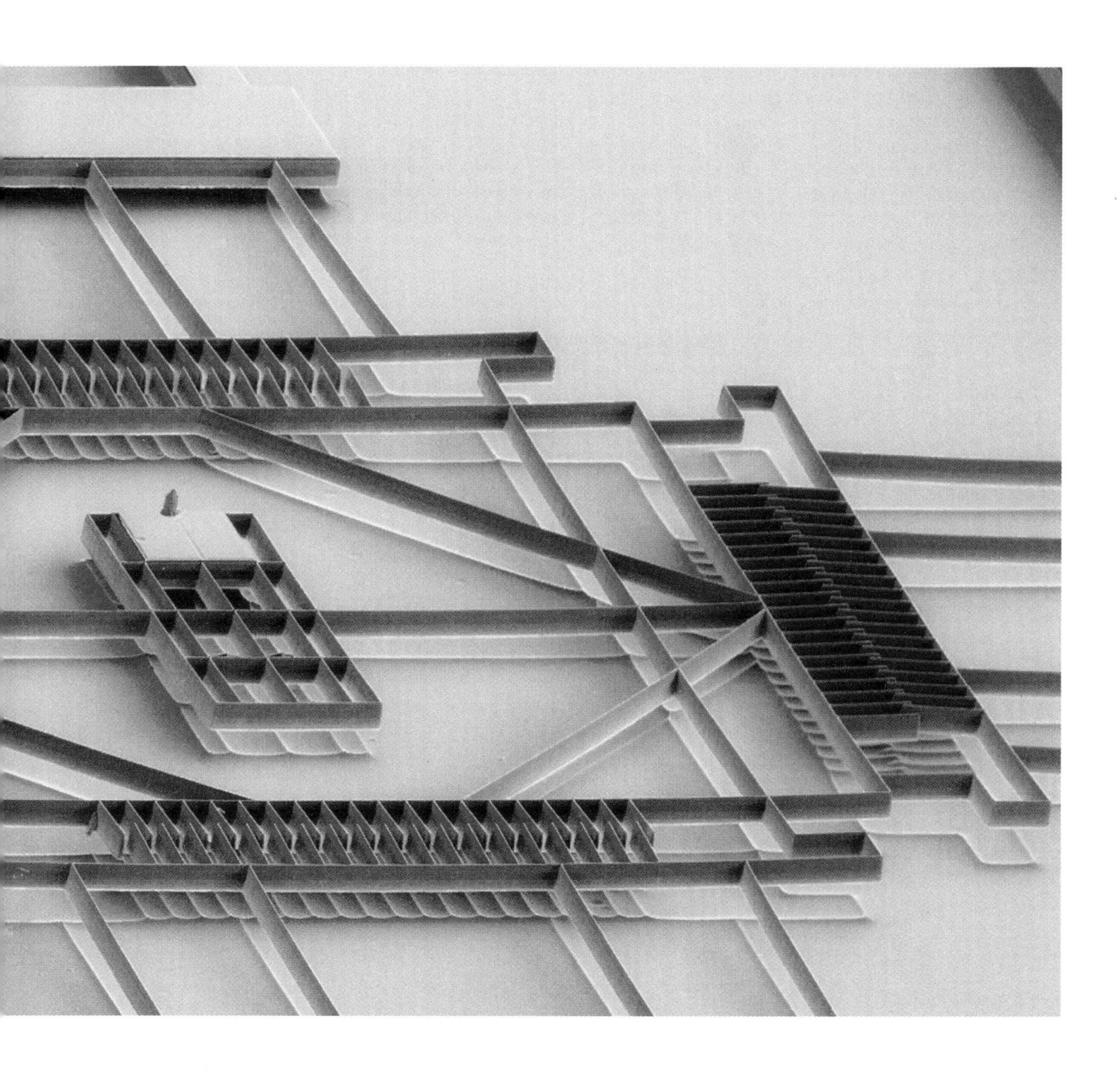

pull on each side of the tip at speeds as high as a million times a second. MacDonald next plans to build an array of STMs.

## The Personal Spectrometer

The Lilliputian infrastructure afforded by MEMS might let chemists and biologists perform their experiments with instruments that fit in the palm of the hand. Westinghouse Science and Technology Center is in the process of reducing to the size of a calculator a 50-pound benchtop spectrometer, used for measuring the mass of atoms or molecules. A miniaturized mass spectrometer presages an era of inexpensive chemical detectors for do-it-yourself toxic monitoring.

In the same vein, Richard M. White, a professor at the University of California at Berkeley, contemplates a chemical factory on a chip. White has begun to fashion millimeter-diameter wells in a silicon chip, each of which holds a different chemical. An electrical voltage causes liquids or powders to move from the wells down a series of channels into a reaction chamber.

These materials are pushed there by micropumps made of piezoelectric materials that constrict and then immediately release sections of the channel. The snakelike undulations create a pumping motion. Once the chemicals are in the chamber, a heating plate causes them to react. An outlet channel from the chamber then pumps out what is produced in the reaction.

A pocket-calculator-size chemical factory could thus reconstitute freeze-dried drugs, perform DNA testing to detect water-borne pathogens or mix chemicals that can then be converted into electrical energy more efficiently than can conventional batteries. MEMS gives microelectronics an opening to the world beyond simply processing and storing information. Automobiles, scientific laboratories, televisions, airplanes and even the home medicine cabinet will never be the same.

# Intelligent Materials

*Inspired by nature, researchers are creating substances that can anticipate failure, repair themselves and adapt to the environment.*

• • •

Craig A. Rogers

Imagine, for a moment, music in your room or car that emanates from the doors, floor or ceiling; ladders that alert us when they are overburdened and may soon collapse under the strain; buildings and bridges that reinforce themselves during earthquakes and seal cracks of their own accord. Like living beings, these systems would alter their structure, account for damage, effect repairs and retire—gracefully, one hopes—when age takes its toll.

Such structures may seem far-fetched. But, in fact, many researchers have demonstrated the feasibility of such "living" materials. To animate an otherwise inert substance, modern-day alchemists enlist a variety of devices: actuators and motors that behave like muscles; sensors that serve as nerves and memory; and communications and computational networks that represent the brain and spinal column. In some respects, the systems have features that can be considered superior to biological functions—some substances can be hard and strong one moment but made to act like Jell-O the next.

These so-called intelligent materials systems have substantial advantages over traditionally engineered constructs. Henry Petroski, in his book *To Engineer Is Human*, perhaps best articulated the traditional principles. A skilled designer always considers the worst-case scenario. As a result, the design contains large margins of safety, such as numerous reinforcements, redundant subunits, backup subsystems and added mass. This approach, of course, demands more natural resources than are generally required and consumes more energy to produce and maintain a structure. It also expends more human effort trying to predict those circumstances under which an engineered artifact will be used and abused.

Trying to anticipate the worst case has a much more serious and obvious flaw, one we read about in the newspapers and hear about on the evening news from time to time: that of being unable to foresee all possible contingencies. Adding insult to injury is the costly litigation that often ensues.

Intelligent materials systems, in contrast, would avoid most of these problems. Made for a given purpose, they would also be able to modify their behavior under dire circumstances. As an example, a ladder that is overloaded with weight could use electrical energy to stiffen and alert the user of the problem. The overload response would be based on the actual life experience of the ladder, to account for aging or damage. As a result, the ladder would be able to evaluate its current health; when it could no longer perform even minimal tasks, the ladder would announce its retirement. In a way, then, the ladder resembles living bone, which remodels itself under changing loads. But unlike bone, which begins to respond within minutes of an impetus but may take months to complete its growth, an intelligent ladder needs to change in less than a second.

## Muscles for Intelligent Systems

Materials that allow structures such as ladders to adapt to their environment are known as actuators. They can change shape, stiffness, position, natural frequency and other mechanical characteristics in response to temperature or electromagnetic fields. The four most common actuator materials being used today are shape-memory alloys, piezoelectric ceramics, magnetostrictive materials and electrorheological and magnetorheological fluids. Although not one of these categories stands as the perfect artificial muscle, each can nonetheless fulfill particular requirements of many tasks.

Shape-memory alloys are metals that at a certain temperature revert back to their original shape after being strained. In the process of returning to their "remembered" shape, the alloys can generate a large force useful for actuation. Most prominent among them perhaps is the family of the nickel-titanium alloys developed at the Naval Ordnance Laboratory (now the Naval Surface Warfare Center). The material, known as Nitinol (Ni for nickel, Ti for titanium and NOL for Naval Ordnance Lab), exhibits substantial resistance to corrosion and fatigue and recovers well from large deformations. Strains that elongate up to 8 percent of the alloy's length can be reversed by heating the alloy, typically with electric current.

Japanese engineers are using Nitinol in micromanipulators and robotics actuators to mimic the smooth motions of human muscles. The controlled force exerted when the Nitinol recovers its shape allows these devices to grasp delicate paper cups filled with water. Nitinol wires embedded in composite materials have also been used to modify vibrational characteristics. They do so by altering the rigidity or state of stress in the structure, thereby shifting the natural frequency of the composite. Thus, the structure would be unlikely to resonate with any external vibrations, a process known to be powerful enough to bring down a bridge. Experiments have shown that embedded Nitinol can apply compensating compression to reduce stress in a structure. Other applications for these actuators include engine mounts and suspensions that control vibration.

The main drawback of shape-memory alloys is their slow rate of change. Because actuation depends on heating and cooling, they respond only as fast as the temperature can shift.

A second kind of actuator, one that addresses the sluggishness of the shape-memory alloys, is based on piezoelectrics. This type of material, discovered in 1880 by French physicists Pierre and Jacques Curie, expands and contracts in response to an applied voltage. Piezoelectric devices do not exert nearly so potent a force as shape-memory alloys; the best of them recover only from less than 1 percent strain. But they act much more quickly, in thousandths of a second. Hence, they are indispensable for precise, high-speed actuation. Optical tracking devices, magnetic heads and adaptive optical systems for robots, ink-jet printers and speakers are some examples of systems that rely on piezoelectrics. Lead zirconate titanate (PZT) is the most widely used type.

Recent research has focused on using PZT actuators to attenuate sound, dampen structural vibrations and control stress. At Virginia Polytechnic Institute and State University, piezoelectric actuators were used in bonded joints to resist the tension near locations that have a high concentration of strain. The experiments extended the fatigue life of some components by more than an order of magnitude.

A third family of actuators is derived from magnetostrictive materials. This group is similar to piezoelectrics except that it responds to magnetic, rather than electric, fields. The magnetic domains in the substance rotate until they line up with an external field. In this way, the domains can expand the material. Terfenol-D, which contains the rare earth element terbium, expands by more than 0.1 percent (see Figure 13.1). This relatively new material has been used in low-frequency, high-power sonar transducers, motors and hydraulic actuators.

**Figure 13.1 MAGNETOSTRICTIVE MATERIAL, such as these Terfenol-D rods that are several centimeters long, changes its length in a magnetic field. Originally developed for military sonar, the alloy is finding use in actuators, vibration controllers and sensors.**

Like Nitinol, Terfenol-D is being investigated for use in the active damping of vibrations.

The fourth kind of actuator for intelligent systems is made of special liquids called electrorheological and magnetorheological fluids. These substances contain micron-size particles that form chains when placed in an electric or magnetic field, resulting in increases in apparent viscosity of up to several orders of magnitude in milliseconds [see "Electrorheological Fluids," by Thomas C. Halsey and James E. Martin; Scientific American, October 1993]. Applications that have been demonstrated with these fluids include tunable dampers, vibration-isolation systems, joints for robotic arms, and frictional devices such as clutches, brakes and resistance controls on exercise equipment. Still, several problems plague these fluids, such as abrasiveness and chemical instability, and much recent work to improve them is aimed at the magnetic substances.

## Nerves of Glass

Providing the actuators with information are the sensors, which describe the physical state of the materials system. Advances in micromachining have created a wealth of promising electromechanical devices that can serve as sensors (see Figures 13.2a-d). I will focus on two types that are well developed now and are the most likely to be incorporated in intelligent systems: optical fibers and piezoelectric materials.

Optical fibers embedded in a "smart" material can provide data in a couple of ways. First, they can simply provide a steady light signal to a sensor; breaks in the light beam indicate a structural flaw that has snapped the fiber. The second, more subtle, approach involves looking at key characteristics of the light—intensity, phase, polarization or a similar feature. The National Aeronautics and Space Administration and other research centers have used such a fiber-optic system to measure the strain in composite materials. Fiber-optic sensors can also measure magnetic fields, deformations, vibrations and acceleration. Resistance to adverse environments and immunity to electrical or magnetic noise are among the advantages of optical sensors.

In addition to serving as actuators, piezoelectric materials make good sensors. Piezoelectric polymers, such as polyvinylidene fluoride (PVDF), are commonly exploited for sensing because they can be formed in thin films and bonded to many kinds of surfaces. The sensitivity of PVDF films to pressure has proved suitable for sensors tactile enough to read Braille and to distinguish grades of sandpaper. Ultrathin PVDF films, perhaps 200 to 300 microns thick, have been proposed for use in robotics. Such a sensor might replicate the capabilities of human skin, detecting temperature and geometric features such as edges and corners or distinguishing between different fabrics.

Actuators and sensors are crucial elements in an intelligent materials system, but the essence of this new design philosophy rests in the manifestation of the most critical of life functions, intelligence. The exact degree of intelligence—the extent to which the material should be smart or merely adaptive—is debatable. At minimum, there must be an ability to learn about the environment and live within it.

The thinking features that the intelligent materials community is trying to create have constraints that the engineering world has never experienced before. Specifically, the tremendous number of sensors, actuators and their associated power sources would argue against feeding all these devices into a

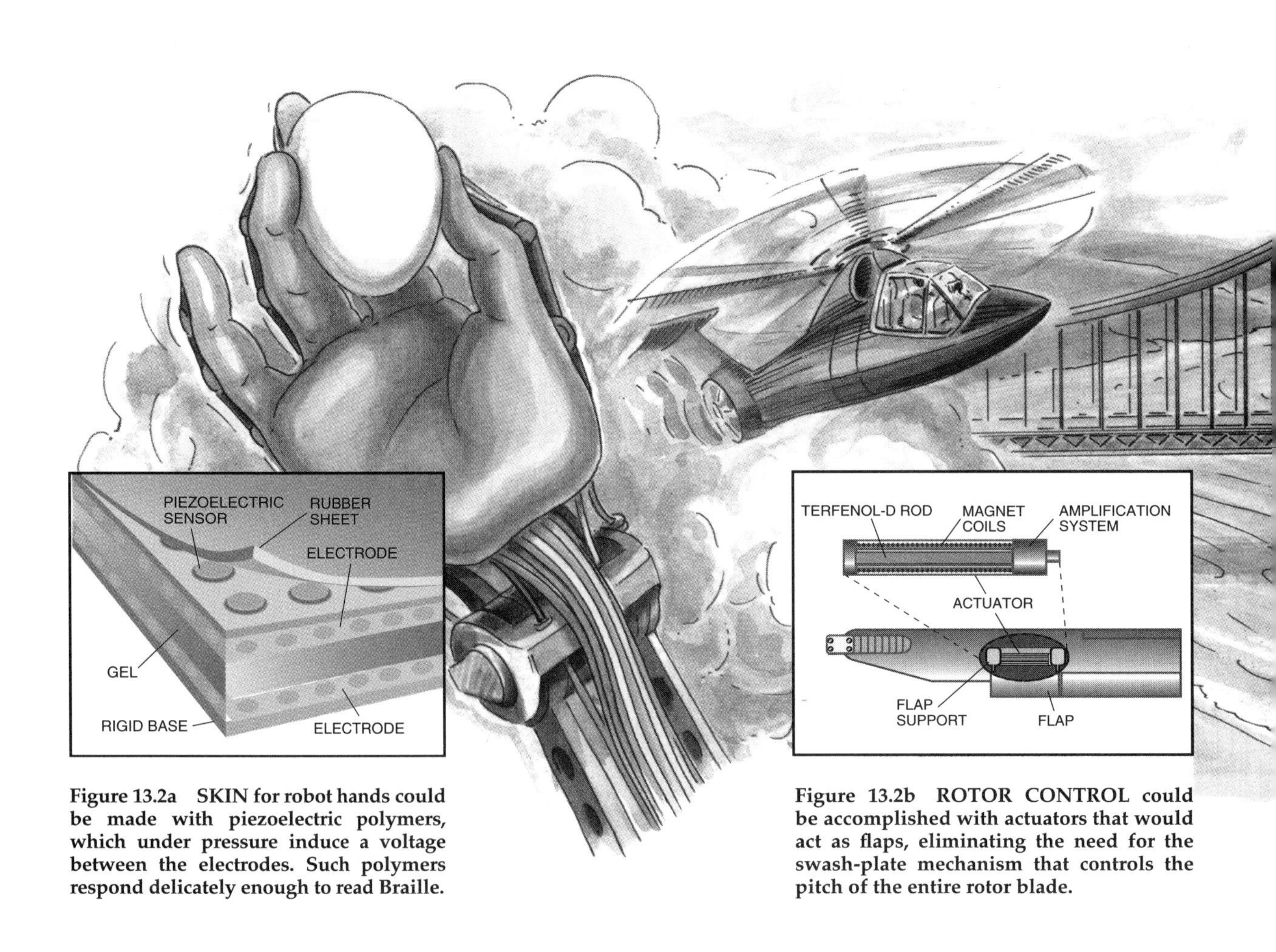


**Figure 13.2a SKIN for robot hands could be made with piezoelectric polymers, which under pressure induce a voltage between the electrodes. Such polymers respond delicately enough to read Braille.**

**Figure 13.2b ROTOR CONTROL could be accomplished with actuators that would act as flaps, eliminating the need for the swash-plate mechanism that controls the pitch of the entire rotor blade.**

central processor. Instead designers have taken clues from nature. Neurons are not nearly so fast as modern-day silicon chips, but they can nonetheless perform complex tasks with amazing speed because they are networked efficiently.

The key appears to be a hierarchical architecture. Signal processing and the resulting action can take place at levels below and far removed from the brain. The reflex of moving your hand away from a hot stove, for example, is organized entirely within the spinal cord. Less automatic behaviors are organized by successively higher centers within the brain. Besides being efficient, such an organization is fault-tolerant: unless there is some underlying organic reason, we rarely experience a burning sensation when holding an iced drink.

The brains behind an intelligent materials system follow a similar organization. In fact, investigators take their cue from research into artificial life, an outgrowth of the cybernetics field. Among the trendiest control concepts is the artificial neural network, which is computer programming that mimics the functions of real neurons. Such software can learn, change in response to contingencies,

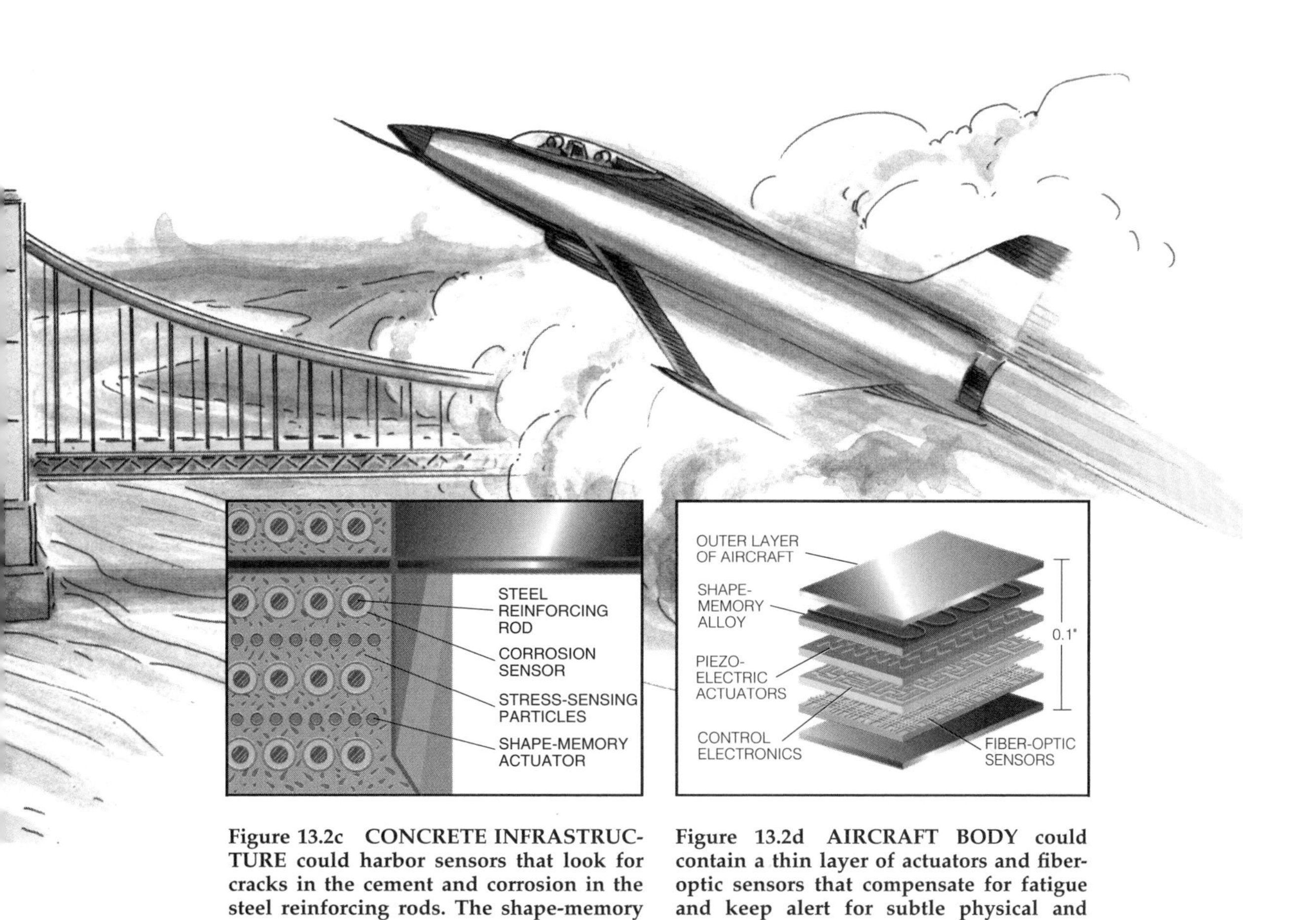


**Figure 13.2c CONCRETE INFRASTRUCTURE could harbor sensors that look for cracks in the cement and corrosion in the steel reinforcing rods. The shape-memory actuators would counteract strain.**

**Figure 13.2d AIRCRAFT BODY could contain a thin layer of actuators and fiber-optic sensors that compensate for fatigue and keep alert for subtle physical and chemical changes presaging failure.**

## Advanced Composites

After they first appeared in the 1960s, advanced composite materials promised a brave new—not to mention light and durable—future. Engineers envisioned buoyant aircraft and automobile bodies that would consume less fuel because they were lighter than traditional vehicles. Although composites have for the most part found their way only into parts in airplanes and automobiles and into sporting equipment, it now looks as though the new class of substance may find its place in a different realm entirely: construction.

The complexity of these materials has a lot to do with their limited appearance in the everyday world. The substances are made of fibers spun from carbon, glass and other materials, which are then fused into a matrix of plastic, ceramic or metal. "They are sophisticated combinations," explains engineer Dick J. Wilkins of the University of Delaware. That intricacy makes mass manufacturing problematic, which has kept their cost high in comparison to the price of wood and metal. "Right now the economics isn't there," Wilkins notes. "You have to have an application that has a high volume."

That application, the composites industry hopes, lies in infrastructure. "In the next five years, there will be much higher use of composites in construction," predicts civil engineer Hota V. S. GangaRao of West Virginia University. He says the materials will initially replace the steel used in reinforcing bars placed in concrete. Composites, which now cost $1.50 to $2.00 a pound, could drop to about $1.20; although this price is several times the 20 to 40 cents per pound of steel, GangaRao calculates, it is still competitive because composites are one fifth the weight of the metal. Eventually, entire structures will be made from them: a few all-composite buildings have already been erected, and plans for bridges are on the drawing boards.

Composites could also be used to repair existing defects. Doug Barnow of the Composites Institute, a trade association based in New York City, points out that the materials patched a severe crack in a bridge along Interstate 95 in Boca Raton, Fla., at a fraction of the cost needed to replace the bridge (the initially planned solution). Composites might also replace some of the thousands of timber piles around New York

anticipate needs and correct mistakes—more than adequate functions for intelligent materials systems. Ultimately, computational hardware and the processing algorithms will determine how complex these systems can become—that is, how many sensors and actuators we can use.

### Brains over Brawn

Engineers are incorporating intelligent materials systems into several areas. NASA uses electroactive materials crafted by researchers at Pennsylvania State University to modify the optics of the *Hubble Space Telescope.*

Perhaps the most mature application at the moment is the control of acoustics. The objective is to reduce sound, be it noise inside an aircraft fuselage shaken by engines or the acoustic signature of a submarine. One way to control noise, of course, is to use brute force. Simply add enough mass to stop the structure from vibrating. In contrast, the intelligent materials approach is to sense the structural oscillations that are radiating the noise and use the actuators distributed throughout the structure to control the most obnoxious vibrations. This concept is the foundation for sound-cancellation headphones used by pilots (and now available through airline gift catalogues), and full systems are being tested on turboprop commuter aircraft.

How far can engineers take intelligent materials? The future lies in developing a system that can behave in more complex ways. For instance, intelligent structures now being demonstrated come with many more sensors than are needed by any one application.

A prospective design might rely on an adaptive architecture in which the sensors can be connected to create the specific system desired. Moreover, the design would be highly flexible. If a particular sensor fails, the adaptive architecture would be able to replace the failed sensor with the next best alterna-

City's waterfront, where harbor cleanup has been so effective that shipworms and other marine borers are now thriving and digesting the wooden structures.

Further inroads may simply depend on better education. "People graduating with engineering degrees are intimidated and don't know how to use [composites]," Wilkins insists. The materials are strong only in the directions in which the fibers run. As a result, they demand more care in design than do, say, metals, which often provide strength in all directions. Unanticipated alterations in load could prove catastrophic for composites. For instance, one speculation about why the carbon-fiber hull of America's Cup racing yacht contender *oneAustralia* snapped was that the crew had extended a sail line to an alternative winch, changing the load on the hull.

Techniques to recycle discarded composite parts have steadily improved in the past several years, another emerging advantage for the materials. At least three different processes have been developed that can separate fibers from resin, Barnow notes, and that ability makes other applications more likely. With more experience and widespread use, engineers should be able to do a lot of good with composites. After all, as Wilkins observes, "the world is not the same stiffness in all directions."

—*Phil Yam, Scientific American*

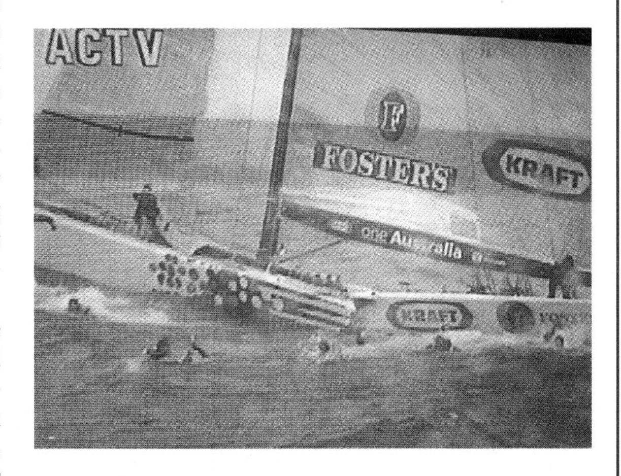


**AW, HULL: the yacht *oneAustralia*, made of composite materials, broke up during an America's Cup race in 1995. Composites can be sensitive to unanticipated changes in load, making them hard to incorporate into designs.**

tive and reconfigure the interconnections and the control algorithm to accommodate this change.

This level of sophistication would clearly tax the manufacturing process. Large arrays of sensors, actuators, power sources and control processors will most likely require three-dimensional interconnections. Such complexity could easily render a smart structure too expensive to build. One solution to forming complex features cheaply is to rely on techniques of computer-chip makers: photolithography. The process, akin to photocopying, can in principle mass-produce components for fractions of a cent per device. A sensor network might therefore resemble the detail of a silicon microchip.

## A New Way of Engineering

Intelligent systems may not only initiate a materials revolution but may also lead to the next step in our understanding of complex physical phenomena. In many ways, they are the ideal recording devices. They can sense their environments, store detailed information about the state of the material over time and "experiment" on the phenomena by changing properties.

The most lasting influence, however, will be on the philosophy of design. Engineers will not have to add mass and cost to ensure safety. They will learn not from the autopsies performed on structures that have failed but from the actual experiences of the edifice. We will soon have the chance to ask structures how they feel, where they hurt, if they have been abused recently. They might even be able to identify the abuser.

Will smart materials systems eliminate all catastrophic failures? Not any more than trees will stop falling in hurricane winds or birds will no longer tumble when they hit glass windows. But intelligent materials systems will enable inanimate objects to become more natural and lifelike. They will be manifestations of the next engineering revolution—the dawn of a new materials age.

## Custom Manufacturing

Where can a cross-country bicyclist find a spare derailleur gear for an Italian bicycle when it breaks down in a small town in Nevada? Ten years or so from now the answer may reside on the floppy disk carried in the rider's backpack. In coming decades, a spare part for a bicycle, automobile or an array of other consumer goods may be simply "printed" from a computer design file at the corner factory, the futuristic equivalent of the all-night copy shop.

In the case of the gear, a machine tool might receive from a file on the disk a geometric description of the broken part. The program could then tell the tool how to deposit a thin layer of structural material by spraying droplets of liquid metal or by directing the energy of a laser onto a bed of metal powder. Subsequent layers deposited in this manner would fuse together, gradually building up a complete derailleur gear.

This approach to manufacturing has its beginnings in a suite of technologies collectively known as rapid prototyping. Today stereolithography, shape deposition, laser sintering and other related technologies can construct full-scale models for testing designs or can help build a tool for making a part. In years to come, improvements to such processes—and an expected decline in equipment costs—may allow them to be used to manufacture finished parts directly or even to fashion new materials.

As they mature, these techniques may also introduce an unprecedented degree of product customization—a machine that could make a bicycle gear one day might make an automobile carburetor the next. This ability to reduce information about an individual's needs to a series of printable computer files is part of a bigger shift away from mass production of a standardized product, as pioneered by auto magnate Henry Ford. Postindustrial manufacturing is evolving toward an era of mass customization: production of substantial quantities of personalized goods. Funding to study this emerging trend—known as agile manufacturing—has come from the Department of Defense. The military is increasingly relying on commercial manufacturers rather than on defense contractors for its needs.

An example of agile manufacturing can already be found in the computer industry. A customer can order a computer over the telephone while choosing from a variety of microprocessors, memory chips, hard disks and monitors. Even more traditional industries, such as those making valves or electrical switching equipment, have begun to follow suit.

Customized fabrication, however, would require more than a trip to the corner factory. Manufacturers need to make more than just the odd bicycle gear. For large-scale fabrication, communications networks may link suppliers to an automobile or blue-jeans maker so that they can fill orders speedily. Networks may also connect customers and factories more closely. A postmillennial Gap store might be equipped with optical scanners that take waist, hip, length and other measurements, send them over a network and have the custom-tailored pants delivered within a matter of days. Clothing stores have begun to experiment with such tailoring, taking the measurements manually. Flexible manufacturing systems—those that let a line of machines be retooled rapidly to make a new type of part or finished good—could enable a factory to respond quickly to changes in demand or to special orders.

A manufacturer must listen not only to customers but also to the competition. Here again, rapid prototyping technology may help. The ability to reproduce a part expeditiously might make

**OIL PUMP MODEL was built with stereolithography, a technique for building up structures layer by layer with a laser.**

it easier to answer the question of how competitors made a product. Computed tomographic images, sometimes superior in resolution to those taken of a patient's heart or brain, already offer General Motors, say, the hypothetical possibility of making a three-dimensional record of a Ford engine block and then building an exact plastic replica using stereolithography—a form of fast prototyping that employs a laser to build up structures made from polymers. This type of reverse engineering may become more common for manufacturing companies as equipment prices fall.

Three-dimensional xerography could also be used for more altruistic purposes. A portable CT scanner might let an image of a damaged bone in a leg be scanned in an ambulance. In the hospital, surgeons would use the model of the bone to plan and perform surgery. Then, while the patient remained in the surgical suite, a prosthesis could be produced and fitted to a limb. The University of Dayton, in fact, has begun to develop a rapid prototyping process for manufacturing ceramic composites, which may be adapted to making prosthetics.

Rapid prototyping techniques can further permit a material to be combined with electronics in unique shapes—the contours of a human cranium, for instance. A technique called shape deposition, which is being explored at Carnegie Mellon and Stanford universities, allows plastics, metal or ceramics to surround microchips and the wires connecting them. The researchers have employed methods that can be used to build wearable computers—they may eventually construct a helmet or hard hat equipped with a working microchip. This chip could store all the information contained in a repair manual; the data can be relayed by an infrared link or a wire to an image projected on a small screen in one lens of a pair of glasses. The helmet could also contain microscopic sensors that ascertain geographic coordinates to within a few feet, useful for hikers or foresters.

Eventually, technology and newfangled management ideas may be combined. "Smart" clothing, for example, may lend itself to agile manufacturing. On a Monday an apparel maker might make shirts that contain sensors to detect the presence of chemical weapons on the battlefield. The next day the same manufacturing line could turn out a garment that dispenses measured quantities of deodorant when sensors in a blouse detect a certain level of perspiration. The true agile manufacturer should be able to handle either sweat or sarin.

—*Gary Stix*, SCIENTIFIC AMERICAN

# High-Temperature Superconductors

*They conduct current without resistance more cheaply than conventional superconductors can and are slowly finding their way to widespread use.*

• • •

Paul C. W. Chu

Nature, it would seem, likes to follow the path of least resistance, be it for heat to transfer, water to flow or a car to travel. If we can follow this path when making and using devices, we can save energy and effort, reduce environmental degradation and, in the long run, improve our standard of living. Unfortunately, nature does not readily reveal the path of least resistance. And it may exist only under certain stringent conditions.

A case in point is the path of zero resistance—superconductivity, the ability to conduct electricity without resistance. Superconductivity was first discovered in 1911, when Dutch physicist Heike Kamerlingh Onnes chilled mercury with liquid helium to four degrees above absolute zero, or four kelvins (a room temperature of 25 degrees Celsius equals 298 kelvins). At that temperature, Onnes observed, mercury would suddenly transmit electricity without loss. Other metals and alloys have since been found to superconduct if cooled to low enough temperatures, most of them to below about 23 kelvins. Such frigid readings—colder than the surface of Pluto—can be reached only with rare gases such as liquefied helium or state-of-the-art refrigeration systems. Despite these conditions, the phenomenon has spawned several technologies—magnetic resonance imaging (MRI) machines, particle accelerators and geological sensors for oil prospecting, among others.

Superconductivity is poised to make an even greater impact on society in the next century, however, thanks to a discovery in the late 1980s. K. Alexander Müller and J. Georg Bednorz of the IBM Research Laboratory in Zurich observed that a ceramiclike substance known as lanthanum barium copper oxide began superconducting at a then record high of 35 kelvins. More dramatic news followed shortly thereafter: in early 1987 Maw-Kuen Wu, then at the University of Alabama at Huntsville, and I, together with our co-workers, demonstrated superconductivity at 93 kelvins in yttrium barium copper oxide, or YBCO for short. At that temperature, YBCO would become superconducting in a bath of liquid nitrogen (see Figure 14.1), which, unlike liquid helium, is abundant and cheap.

That work sparked a flurry of activity as researchers sought other superconducting cuprates, as these copper oxide compounds are called. Indeed, physicists have discovered more than 100 superconductors with critical temperatures that exceed those of the best conventional superconductors. (This fact prompted some theorists to plead, "Stop discovering more new ones before we understand what we have!")

The novel materials raised many questions, perhaps foremost among them: Can superconductors follow in the footsteps of their cousins, the semiconductors, and dramatically change our lives for the better? A qualified "yes" is not overly optimistic, because superconductors can touch every aspect of our existence that involves electricity. Superconducting trains, nearly perfect, large energy storage systems and ultrafast computers are not realistic goals at the moment, but plenty of other applications are, in principle, possible soon: efficient generation, transmission and storage of electricity; detection of electromagnetic signals too small to be sensed by conventional means; protection of electrical grids from power surges, sags and outages; and the development of faster and more compact cellular communications technology.

## A Troublesome Material

Although they may sound rather mundane, these potential uses are in a way almost too good to be true, considering the myriad hurdles that became apparent shortly after the discovery of the cuprates. One of the biggest was that cuprates carried only a limited amount of electricity without resistance, a problem stemming from the positioning of the layers that made up the materials. If the layers did not line up properly, electrons would bump into the boundary in the misaligned region and slow down. Magnetic fields further exacerbated the situation, as they could easily penetrate this misaligned region and disrupt the free flow of current. In fact, even a

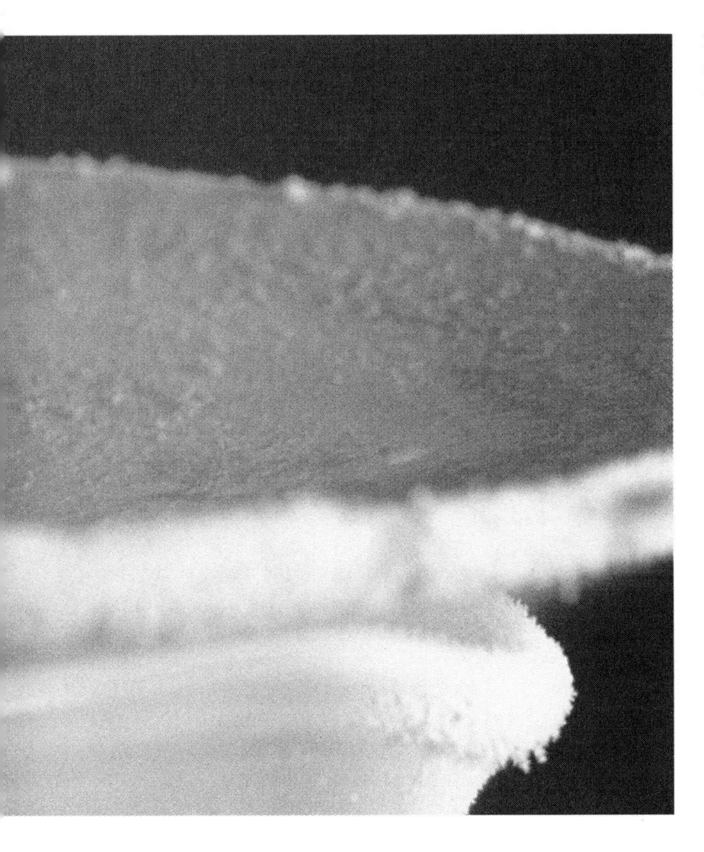

**Figure 14.1 KEPT IN SUSPENSE: a magnet floats above a superconductor cooled with liquid nitrogen, which repels all external magnetic fields.**

perfectly aligned material can fall victim to this intrusion if the magnetic field is extremely strong.

Researchers found one way around these hindrances: lay down micron-thin layers of the material on well-organized substrates. The process had the effect of lining up the superconducting layers more accurately. Although thin films do not carry tremendous amounts of current, many organizations have begun marketing instruments based on them. Du Pont, the Massachusetts Institute of Technology's Lincoln Laboratory, Conductus, Illinois Superconductor, and Superconductor Technologies Incorporated (STI) are all making devices that operate in the microwave frequencies for military instruments and cellular telephone systems. The superconducting films provide greater signal strength and process signals more efficiently in a smaller package than can ordinary conductors. Conductus and IBM are also making magnetic-field sensors known as superconducting quantum interference devices (see Figure 14.2), or SQUIDs [see "SQUIDs," by John Clarke; SCIENTIFIC AMERICAN, August 1994]. These devices perform as well at the liquid nitrogen temperature of 77 kelvins as do conventional SQUIDs operating at 4.2 kelvins. Conductus currently sells models for educational and research purposes.

While some investigators traveled the thin-film route, others tackled the intractable problem of limited current capacity and intrusive magnetic fields head-on, in the hopes of having wires and motors and other "bulk" applications. They devised many ways to surmount the obstacles. For instance, careful processing that aligned the layers of the cuprates boosted the current capacity. Investigators also sought to introduce structural defects into selected parts of the superconductor, which would act to "pin down" magnetic fields and limit their disruptive tendency.

Such modifications have produced remarkable results. The maximum current density YBCO can carry is now one million amperes per square centimeter at 77 kelvins, dropping only to 400,000 amperes when a magnetic field of nine teslas is applied. Both values are much higher than initial results, when YBCO could carry only 10 amperes per square centimeter and lost all conductivity in only a 0.01-tesla field. In many respects, the current capacity now obtainable is comparable to those of conventional superconductors. When cooled to identical temperatures and placed in high fields, the cuprates in some ways outperform their low-temperature cousins.

Still, bulk applications faced another hurdle. The cuprates are essentially ceramics, which are brittle and difficult to form into wires. Through new processing techniques and materials selection, researchers have managed to coax flexible wires out of the breakable substance (see Figure 14.3). They pack a precursor powder into a silver tube that is rolled and pressed into wires. Subsequent baking converts the powder into a bismuth-based cuprate. Short samples can carry 200,000 amperes per square centimeter at 4.2 kelvins (about 200 times the amount copper can usually handle) and 35,000 amperes at 77 kelvins. American Superconductor can now routinely spin out kilometer-long lengths of wire. By using ion beams, physicists at Los Alamos National Laboratory recently produced samples of flexible YBCO tape that can resist magnetic fields much better than bismuth wires do.

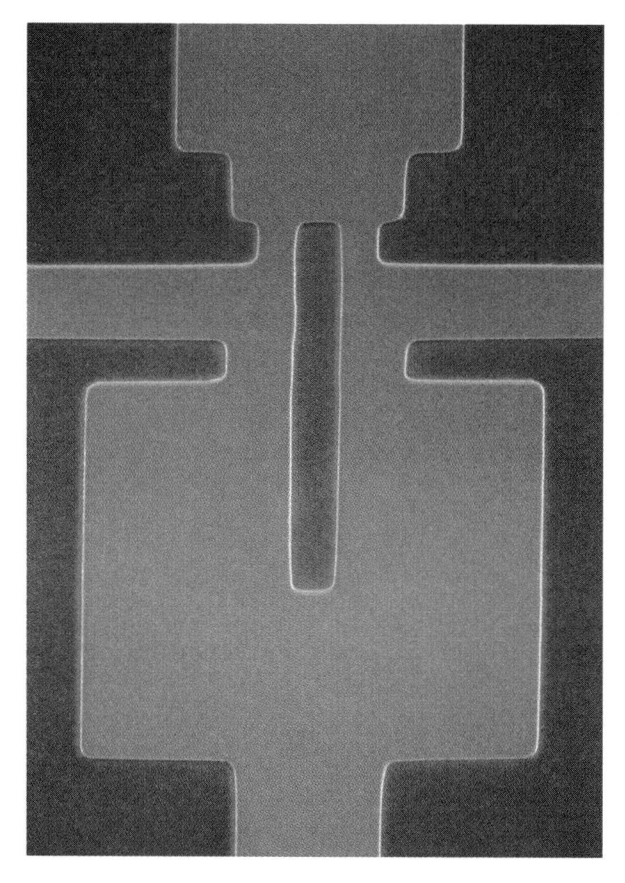

**Figure 14.2 SQUID, or superconducting quantum interference device, serves as a highly sensitive detector of magnetic fields. This one, only 30 microns across, contains two Josephson junctions (*not visible*) lying just above the horizontal strip that runs across the image.**

Several devices demonstrating the feasibility of bulk applications have been constructed. Intermagnetics General and the Texas Center for Superconductivity at the University of Houston have built different types of cuprate magnets that can generate up to two teslas, about five times the field provided by the best permanent magnet. Reliance Electric will use American Superconductor's wire to produce a five-horsepower motor. These and other institutions have also crafted flywheels to store energy and fault-current limiters to shunt electrical surges from equipment. Although some of these devices have analogues among ordinary conductors, as superconducting devices they should perform with greater efficiency and capacity.

## Prototypes to Market?

Predicting the future is always a bit hazardous, even more so in the absence of an adequate present. Nevertheless, I will venture a few prognostications about the impact of high-temperature superconductivity on our lives in the next 10 to 30 years, based on the developments since 1987.

Many of the demonstration devices now being built will become ubiquitous, as manufacturing and processing become more refined and performance improves. SQUIDs, which can detect the weak magnetic signals from the heart and brain, will become a common tool for the noninvasive diagnosis of diseases. Tests have shown that these sensors can pinpoint the areas of the brain responsible for focal epilepsy. SQUIDs will also become standard issue in nondestructive testing of infrastructure such as oil pipes and bridges, because fatigued metal produces a unique magnetic signature. The advantages of liquid nitrogen should render these detectors more widely used in all areas of scientific investigation.

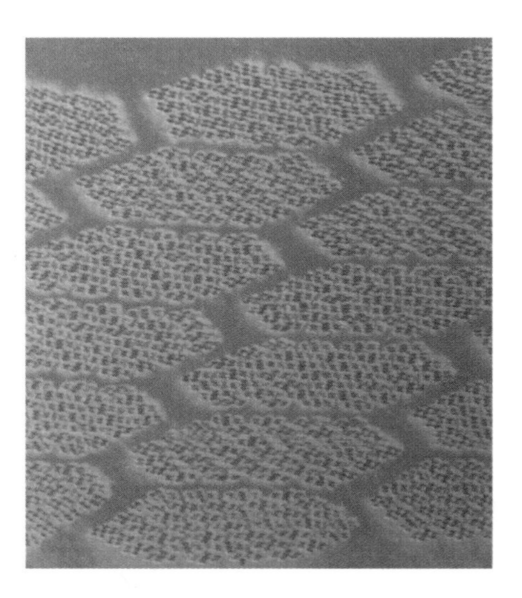

**Figure 14.3 SUPERCONDUCTING WIRE cut laterally reveals filaments four microns thick that are packed into hexagonal patterns. This design, used by American Superconductor, helps to make the brittle ceramic bendable and strong.**

MRI devices will probably become smaller and more efficient. More sensitive superconducting amplifiers and coil detectors will mean that the magnetic field required for imaging can be weaker, which would result in a smaller and cheaper machine. The greater sensitivity will lead to faster signal processing, hence greatly lowering the present cost of operating the machine.

Less visible but important economically, high-temperature superconductors will work their way into microwave communications systems, acting as filters and antennas. They will prove indispensable in boosting the capacity of cellular telephone base stations by a factor of three to 10. They will also be common equipment in military aircraft to filter out extraneous radar signals that could confuse the onboard computers.

Highly populated areas may see old underground power cables replaced with superconducting lines to meet the increasing demand for electricity. Such transmission lines may also reduce electric rates—about 15 percent of the bill stems from loss caused by electrical resistance. Power stations will rely on the materials for current limiters, providing more stable voltages for a computer-dependent society.

Energy storage is also a strong possibility. Superconducting magnetic energy storage (SMES) devices may become widespread. Essentially, a superconducting coil would be charged and then wound into a circle. The current will theoretically flow without loss. When the electricity is needed, the coil is snapped back into the main grid, providing a boost of electricity. Prototype SMES devices, using low-temperature superconductors, are now being tested. Advanced flywheels, which would be supported by frictionless superconducting bearings and would spin continuously until tapped for their energy, would serve a similar purpose.

The cuprates may also prove economically feasible in equipment for space exploration. Away from the direct rays of the sun, the temperatures in space are below that needed to sustain superconductivity for many of these materials. With that realization in mind, the National Aeronautics and Space Administration has funded the development of prototype sensing and electromechanical devices for spaceship use.

Some researchers are exploring even more remote applications, specifically in computer technology. One is to make Josephson junction circuits. A Josephson junction, crafted by sandwiching a thin insulating barrier between two superconducting layers, can be made to turn on and off rapidly with low power. The junctions could replace the circuits in computers and in theory boost the speed of computation by up to 50 times. Technical obstacles, however, have prevented significant progress toward an all-superconducting computer. A hybrid system may be viable. The key problems here are making reliable superconducting circuits and designing appropriate interfaces between superconductors and semiconductors, not to mention fighting off the competition from the ever improving field of semiconductors.

## Superconducting Secrets

Even greater technological change may rely on advances in basic research of the superconductors. The complexity of the material has made the mechanism behind high-temperature superconductivity impervious to probing. The traditional theory of superconductivity states that vibrations of the solid cause electrons, which ordinarily repel one another, to form pairs. These pairs can then race along without resistance.

This conception, however, appears inadequate for cuprates. The high transition temperature means that the solid would have to shake so much that the lattice structure of the compound would not be stable enough for electron pairs to form. Something else must be matching up the electrons. One clue lies in the normal (that is, nonsuperconducting) state. Here the materials show unusual electric and magnetic properties that defy prevailing wisdom. Many experiments are being conducted to narrow the field of theories. I suspect that many mechanisms are acting together to produce superconductivity in cuprates and that they will be elucidated by the year 2005.

Once the materials are understood, even higher transition temperatures may be reached. The confirmed mark for a substance under normal conditions is 134 kelvins, first observed in 1993 by Andreas Schilling and his colleagues at the Swiss Federal Institute of Technology in Zurich in mercury barium calcium copper oxide (see Figure 14.4). By squeezing the compound, Dave Mao of Geophysical Lab and I, along with our co-workers, raised the critical temperature to 164 kelvins. Such a temperature, equal to –109 degrees Celsius, is attainable with technology used in household air-conditioning.

In fact, a room-temperature superconductor may be found; most theories do not exclude the possibility. Sporadic but irreproducible results have appeared suggesting superconductivity as high as 250 kelvins (–23 degrees C). A room-temperature superconductor would surely initiate another industrial revolution. Although the pace of improvement has made workers optimistic, the existence of a technology alone does not guarantee it a major position in a market-oriented society. The cost-benefit factor dictates the outcome. Hence, the challenge is to reduce the price to process the material, fabricate the device and implement the technology.

From 1987 to 1995, scientists have made the normal abnormal by discovering high-temperature superconductors. Then they have made the abnormal normal by unraveling some mysteries of the phenomenon. Now they are trying to make the normal practical by demonstrating the technical feasibility of the effect. Although unforeseen applications are certain to arise—no one predicted that MRI technology would emerge from superconductors—the high-temperature wonderland will most likely consist of subtle yet economically profound changes, a conversion of esoteric technology into instruments we can rely on every day.

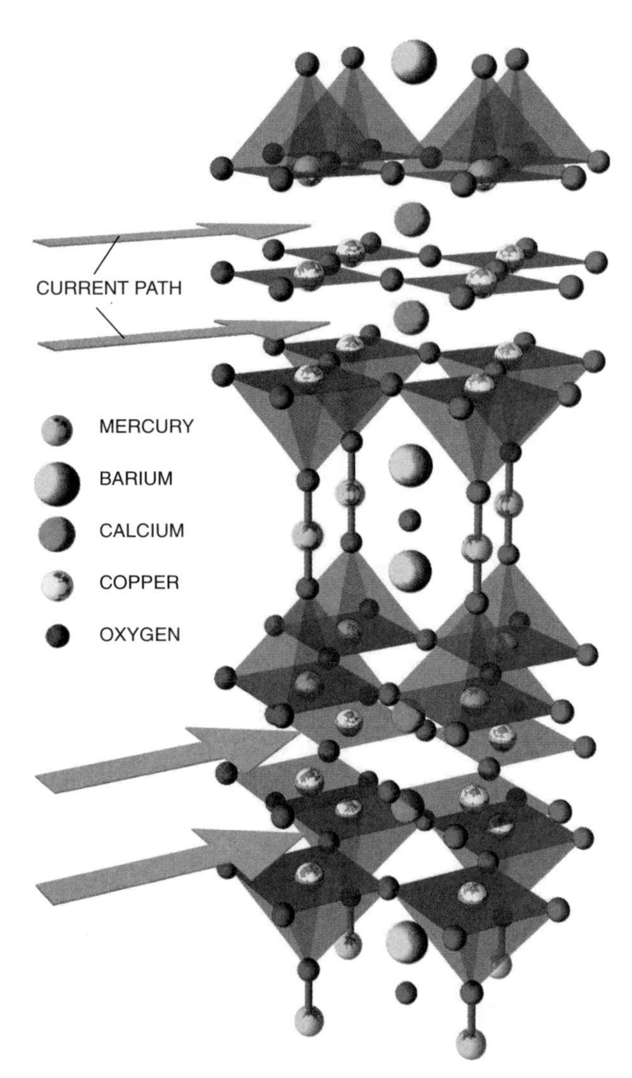


**Figure 14.4 HIGHWAYS FOR ELECTRONS are located between planes of copper and oxygen atoms, as shown in this representation of a mercury barium calcium copper oxide, which loses its electrical resistance at 134 kelvins—among the highest of the high-temperature superconductors.**

# Commentary: Robotics in the 21st Century

*Automatons may soon find work as subservient household help.*

• • •

Joseph F. Engelberger

Since Unimation, Inc., installed the first industrial robot in 1961 to unload parts from a die-casting operation, more than 500,000 similar constructs have gone to work in factories around the world. They are common sights in chemical processing plants, automobile assembly lines and electronics manufacturing facilities, replacing human labor in repetitive and possibly dangerous operations. But how will robotics evolve in the immediate decades ahead? Can it move from the industrial setting to serve people in their daily lives?

In his story "There Will Come Soft Rains," Ray Bradbury forecast that our homes would become enveloping automaton systems that could outlast the human inhabitants. Isaac Asimov came to different conclusions. In his robot novels, Asimov envisioned stand-alone robots that would serve and mingle with humans to our everlasting mutual benefit. The answer can be gleaned from real-world experience coupled with the speculations of these two science-fiction giants.

Finding first for Bradbury, we already have the Smart House project, an (expensive) option for home buyers in which a central computer optimizes heat, light, air conditioning and security. Add automatic control of communications, entertainment, data seeking and shopping in cyberspace, and seemingly much of Bradbury's conjecture is justified.

But not completely. Bradbury's house went further, offering automated cooking, cleaning and personal hygiene. A slew of little robot mice, for instance, would dart out from the baseboards throughout the house to pick up dirt. Human occupants literally did not have to lift a finger.

Such physical intervention, however, is where Bradbury's vision loses some credibility. Although we already have automation in washing machines, dishwashers, coffeemakers and so forth, these devices are loaded manually, with human hands. Robotic mice that do away with dust might be technically feasible at great cost, but it seems more practical to push around a vacuum cleaner every so often.

In that regard, I think Asimov's scenario becomes more likely. Rather than a specialized device, a household robot would be a stand-alone automaton.

It would do the chores just as we do, using the same equipment and similar tools and responding to spoken commands and supplying verbal reports.

Therefore, the robots that serve us personally in the near future will of necessity be somewhat anthropomorphic, just as Asimov envisioned. To share a household with a human, the robot must be able to travel autonomously throughout the living quarters, to see and interpret needs, and to provide materials and services with a gentle and precise touch.

It may be that before the early decades of the 21st century become history, some profound invention will alter robotics. But it behooves this would-be oracle to stay within the bounds of current technology and logical extensions thereof. And that is not much of a constraint! Roboticists have a substantial toolbox in hand today—low-cost electronics, servomechanisms, controllers, sensors and communications equipment, to name just a few categories. Moreover, these instruments of construction are steadily evolving, particularly those in sensory perception. Active and passive beacons, stereo vision and even a receiver for the Global Positioning System (a network of satellites that broadcasts positional information) will enable a robot to navigate its environment effortlessly. Voice synthesis and recognition will ensure understanding of the human overseer's needs. Safety precautions, such as rules similar to Asimov's three laws of robotics—which can be loosely paraphrased as "Protect humans; obey humans; protect yourself"—can readily be embedded.

A robot built with these tools will probably first see duty as a companion to the elderly and handicapped, perhaps before the end of this century. It would give ambulatory aid (offer an arm), fetch and carry, cook, clean, monitor vital signs, entertain, and communicate with human caregivers located elsewhere. This is not science fiction. We have the capability now—solid engineering is all that is required.

Still, I point out that the turn-of-the-century household robot, however useful, would be a far cry from a loyal staff of butler, cook, maid and nurse. Human intervention would still be necessary for personal hygiene, dressing, grooming and invasive administration of medicine. And artificial intelligence would not provide fully satisfying conversational gambits.

Personal-service robots may also encourage advances in their industrial cousins, which are still consigned largely to "put and take" actions. Making robots more adaptive and communicative may help reduce the preprogramming now required. As robots improve, they will be assigned to more hot, hazardous and tedious tasks. I fondly hope that sooner, rather than later, the National Aeronautics and Space Administration and other space agencies worldwide will realize that robots should be our emissaries in space. Flesh and blood do not belong in that desperately hostile environment. Using robots may ultimately make the human colonization of space proceed much faster. Robots on Mars, for instance, could create a friendly environment before humans arrive.

In the course of terrestrial evolution, humanity is only a recent development. Our continuing advance is inhibited by the ponderous nature of natural selection and by the laboriousness of the learning processes that endow our progeny with their forebears' wisdom. In contrast, every new robot can rapidly incorporate the best physical and intellectual capacity available at the time. Within seconds, all prior robotic experience can be downloaded. Robotics may very well determine how human activity evolves in the 21st century.

# PART V
## ENERGY AND ENVIRONMENT

# Solar Energy

*Technology will allow radiation from the sun to provide nonpolluting and cheap fuels, as well as electricity.*

• • •

William Hoagland

Every year the earth's surface receives about 10 times as much energy from sunlight as is contained in all the known reserves of coal, oil, natural gas and uranium combined. This energy equals 15,000 times the world's annual consumption by humans. People have been burning wood and other forms of biomass for thousands of years, and that is one way of tapping solar energy. But the sun also provides hydropower, wind power and fossil fuels—in fact, all forms of energy other than nuclear, geothermal and tidal.

Attempts to collect the direct energy of the sun are not new. In 1861 a mathematics instructor named Augustin-Bernard Mouchot of the Lycée de Tours in France obtained the first patent for a solar-powered motor. Other pioneers also investigated using the sun's energy, but the convenience of coal and oil was overwhelming. As a result, solar power was mostly forgotten until the energy crisis of the 1970s threatened many major economies.

Economic growth depends on energy use. By 2025 the worldwide demand for fuel is projected to increase by 30 percent and that for electricity by 265 percent. Even with more efficient use and conservation, new sources of energy will be required. Solar energy could provide 60 percent of the electricity and as much as 40 percent of the fuel.

Extensive use of more sophisticated solar energy technology will have a beneficial impact on air pollution and global climatic change. In developing countries, it can alleviate the environmental damage caused by the often inefficient practice of burning plant material for cooking and heating. Advanced solar technologies have the potential to use less land than does biomass cultivation: photosynthesis typically captures less than 1 percent of the available sunlight, but modern solar technologies can, at least in the laboratory, achieve efficiencies of 20 to 30 percent. With such efficiencies, the U.S. could meet its current demand for energy by devoting less than 2 percent of its land area to energy collection.

It is unlikely that a single solar technology will predominate (see Figure 15.1). Regional variations in economics and the availability of sunlight will naturally favor some approaches over others. Electricity may be generated by burning biomass, erecting wind turbines, building solar-powered heat engines, laying out photovoltaic cells or harnessing the energy in rivers with dams. Hydrogen fuel can be produced by electrochemical cells or biological processes—involving microorganisms or enzymes—that are driven by sunlight. Fuels such as ethanol and methanol may be generated from biomass or other solar technologies.

Solar energy also exists in the oceans as waves and gradients of temperature and salinity, and they, too, are potential reservoirs to tap. Unfortunately, although the energy stored is enormous, it is diffuse and expensive to extract.

## Growing Energy

Agricultural or industrial wastes such as wood chips can be burned to generate steam for turbines. Such facilities are competitive with conventional electricity production wherever biomass is cheap. Many such plants already exist, and more are being commissioned. Recently in Värnamo, Sweden, a modern power plant using gasified wood to fuel a jet engine was completed. The facility converts 80 percent of the energy in the wood to provide six megawatts of power and nine megawatts of heat for the town. Although biomass combustion can be polluting, such technology makes it extremely clean.

Progress in combustion engineering and biotechnology has also made it economical to convert plant material into liquid or gaseous fuels. Forest products, "energy crops," agricultural residues and other wastes can be gasified and used to synthesize methanol. Ethanol is released when sugars, derived from sugarcane or various kinds of grain crops or from wood (by converting cellulose), are fermented.

Alcohols are now being blended with gasoline to enhance the efficiency of combustion in car engines and to reduce harmful tail-pipe emissions. But ethanol can be an effective fuel in its own right, as researchers in Brazil have demonstrated. It may be cost-competitive with gasoline by 2000. In the future, biomass plantations could allow such energy to be "grown" on degraded land in developing nations. Energy crops could also allow for better land management and higher profits. But much research is needed to achieve consistently high crop yields in diverse climates.

Questions do remain as to how useful biomass can be, even with technological innovations. Photosynthesis is inherently inefficient and requires large supplies of water. A 1992 study commissioned by the United Nations concluded that 55 percent of the world's energy needs could be met by biomass by 2050. But the reality will hinge on what other options are available (see Figure 15.2).

## Wind Power

Roughly 0.25 percent of the sun's energy reaching the lower atmosphere is transformed into wind—a minuscule part of the total but still a significant source of energy. By one estimate, 80 percent of the electrical consumption in the U.S. could be met by the wind energy of North and South Dakota alone. The early problems surrounding the reliability of "wind farms" have now been by and large

**Figure 15.1 DIVERSE DEVICES aid in capturing solar energy. Wind turbines (*a*) draw out the energy stored in the atmosphere through differential heating by the sun. A solar furnace (*b*) uses radiation reflected onto a central tower to drive an engine. Solar panels (*c* and *background*) employ photovoltaic cells to create electricity. And crops such as sugarcane (*d*) tap sunlight by photosynthesis. The sugar can be converted to alcohol, a clean fuel.**

c

d

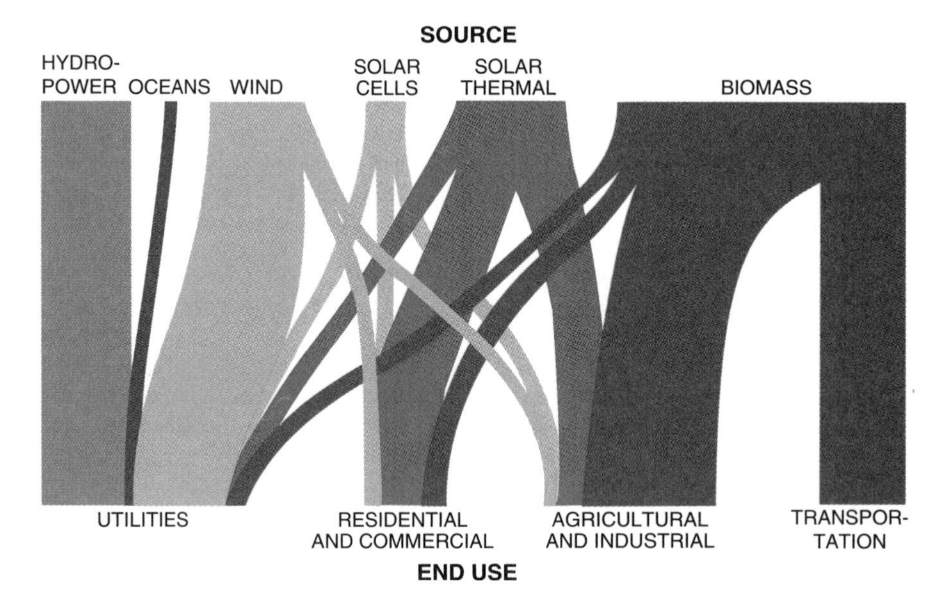


**Figure 15.2 DISTRIBUTION of renewable solar energy projected for the year 2000 shows that many different means of tapping the resource will play a role.**

resolved, and in certain locations the electricity produced is already cost-competitive with conventional generation.

In areas of strong wind—an average of more than 7.5 meters per second—electricity from wind farms costs as little as $0.04 per kilowatt-hour. The cost should drop to below $0.03 per kilowatt-hour by the year 2000. In California and Denmark more than 17,000 wind turbines have been completely integrated into the utility grid. Wind now supplies about 1 percent of California's electricity.

One reason for the reduction is that stronger and lighter materials for the blades have allowed wind machines to become substantially larger. The turbines now provide as much as 0.5 megawatt apiece. Advances in variable-speed turbines have reduced stress and fatigue in the moving parts, thus improving reliability. Over the next 20 years better materials for air foils and transmissions and smoother controls and electronics for handling high levels of electrical power should become available.

One early use of wind energy will most likely be for islands or other areas that are far from an electrical grid. Many such communities currently import diesel for generating power, and some are actively seeking alternatives. By the middle of the next century, wind power could meet 10 to 20 percent of the world's demand for electrical energy.

The major limitation of wind energy is that it is intermittent. If wind power constitutes more than 25 to 45 percent of the total power supply, any shortfall causes severe economic penalties. Better means of energy storage would allow the percentage of wind power used in the grid to increase substantially.

## Heat Engines

One way of generating electricity is to drive an engine with the sun's radiant heat and light. Such solar-thermal electric devices have four basic components, namely, a system for collecting sunlight, a receiver for absorbing it, a thermal storage device and a converter for changing the heat to electricity. The collectors come in three basic configurations: a parabolic dish that focuses light to a point, a parabolic trough that focuses light to a line and an array of flat mirrors spread over several acres that reflect light onto a single central tower.

These devices convert between 10 and 30 percent of the direct sunlight to electricity. But uncertainties remain regarding their life span and reliability. A particular technical challenge is to develop a Stirling engine that performs well at low cost. (A Stirling engine is one in which heat is added continuously from the outside to a gas contained in a closed system.)

Solar ponds, another solar-thermal source, contain highly saline water near their bottom. Typically, hot water rises to the surface, where it cools off. But salinity makes the water dense, so that hot water can stay at the bottom and thus retain its heat. The pond traps the sun's radiant heat, creating a high temperature gradient. Hot, salty fluid is drawn out from the bottom of the pond and allowed to evaporate; the vapor is used to drive a Rankine-cycle engine similar to that installed in cars. The cool liquid at the top of the pond can also be used, for air-conditioning.

A by-product of this process is freshwater from the steam. Solar ponds are limited by the large amounts of water they need and are more suited to remote communities that require freshwater as well as energy. Use of solar ponds has been widely investigated in countries with hot, dry climates, such as in Israel.

## Solar Cells

The conversion of light directly to electricity, by the photovoltaic effect, was first observed by the French physicist Edmond Becquerel in 1839. When photons shine on a photovoltaic device, commonly made of silicon, they eject electrons from their stable positions, allowing them to move freely through the material. A voltage can then be generated using a semiconductor junction. A method of producing extremely pure crystalline silicon for photovoltaic cells with high voltages and efficiencies was developed in the 1940s. It proved to be a tremendous boost for the industry. In 1958 photovoltaics were first used by the American space program to power the radio of the *U.S. Vanguard I* space satellite with less than one watt of electricity.

Although significant advances have been made since 1975—the current record for photovoltaic efficiency is more than 30 percent—cost remains a barrier to widespread use. There are two approaches to reducing the high price: producing cheap materials for so-called flat-plate systems, and using lenses or reflectors to concentrate sunlight onto smaller areas of (expensive) solar cells. Concentrating systems must track the sun and do not use the diffuse light caused by cloud cover as efficiently as flat-plate systems. They do, however, capture more light early and late in the day.

Virtually all photovoltaic devices operating today are flat-plate systems. Some rotate to track the sun, but most have no moving parts. One may be optimistic about the future of these devices because commercially available efficiencies are well below theoretical limits and because modern manufacturing techniques are only now being applied. Photovoltaic electricity produced by either means should soon cost less than $0.10 cents per kilowatt-hour, becoming competitive with conventional generation early in the next century.

## Storing Energy

Sunlight, wind and hydropower all vary intermittently, seasonally and even daily. Demand for energy fluctuates as well; matching supply and demand can be accomplished only with storage. A study by the Department of Energy estimated that by 2030 in the U.S., the availability of appropriate storage could enhance the contribution of renewable energy by about 18 quadrillion British thermal units per year.

With the exception of biomass, the more promising long-term solar systems are designed to produce only electricity. Electricity is the energy carrier of choice for most stationary applications, such as heating, cooling, lighting and machinery. But it is not easily stored in suitable quantities. For use in transportation, lightweight, high-capacity energy storage is needed.

Sunlight can also be used to produce hydrogen fuel. The technologies required to do so directly (without generating electricity first) are in the very early stages of development but in the long term may prove the best. Sunlight falling on an electrode can produce an electric current to split water into hydrogen and oxygen, by a process called photoelectrolysis. The term "photobiology" is used to describe a whole class of biological systems that produce hydrogen. Even longer-term research may lead to photocatalysts that allow sunlight to split water directly into its component substances.

When the resulting hydrogen is burned as a fuel or is used to produce electricity in a fuel cell, the only by-product is water. Apart from being environmentally benign, hydrogen provides a way to

## A New Chance for Solar Energy

Solar power is getting cheaper—in fact, the cost of filching the sun's rays has fallen more than 65 percent since 1985. It has not become inexpensive enough, though, to rival fossil fuels, so solar energy remains a promising, not yet fully mature alternative. Sales run only about $1 billion annually, as opposed to roughly $800 billion for standard sources, and solar customers still generally reside in isolated areas, far from power grids.

But a new proposal from an American utility may well make solar power conventional—or at least more competitive. Enron Corporation, the largest U.S. supplier of natural gas, recently joined forces with Amoco Corporation, owner of the photovoltaic cell producer Solarex. The two companies intend to build a 100-megawatt solar plant in the Nevada desert by the end of 1996. The facility, which could supply a city of 100,000, will initially sell energy for 5.5 cents a kilowatt-hour—about three cents cheaper on average than the electricity generated by oil, coal or gas. "If they can pull this off, it can revolutionize the whole industry," comments Robert H. Williams of Princeton University. "If they fail, it is going to set back the technology 10 years."

Despite its magnitude, the $150-million plan does not mean that the solar age has finally dawned: Enron's low price is predicated on tax exemptions from the Department of Energy and on guaranteed purchases by the federal government. Nor does it mark a sudden technological breakthrough: Solarex manufactures a conventional thin-film, silicon-based photovoltaic cell that is able to transform into electricity about 8 percent of the sunlight that reaches it. Rather the significance of Enron's venture—should the bid be accepted by the government—is that it paves the way for other companies to make large-scale investments in solar power.

Such investments could bring the price of solar-power technology and delivery down even further—for both large, grid-based markets and for the more dispersed, off-the-grid markets that are the norm in many developing countries. "This marks a shift in approach," explains Nicholas Lenssen, formerly at the Worldwatch Institute in Washington, D.C., and now at E Source in Boulder, Colo. "It allows them to attract lower-risk, long-term capital, not just venture capital, which is very costly." Which all means the Nevada desert may soon be home to a very different, but still very hot, kind of test site.

—*Marguerite Holloway, Scientific American*

alleviate the problem of storing solar energy. It can be held efficiently for as long as required. Over distances of more than 1,000 kilometers, it costs less to transport hydrogen than to transmit electricity. Residents of the Aleutian Islands have developed plans to make electricity from wind turbines, converting it to hydrogen for storage. In addition, improvements in fuel cells have allowed a number of highly efficient, nonpolluting uses of hydrogen to be developed, such as electric vehicles powered by hydrogen.

A radical shift in our energy economy will require alterations in the infrastructure. When the decision to change is made will depend on the importance placed on the environment, energy security or other considerations. In the U.S., federal programs for research into renewable energy have been on a roller-coaster ride. Even the fate of the Department of Energy is uncertain.

At present, developed nations consume at least 10 times the energy per person than is used in developing countries. But the demand for energy is rising fast everywhere. Solar technologies could enable the developing world to skip a generation of infrastructure and move directly to a source of energy that does not contribute to global warming or otherwise degrade the environment. Developed countries could also benefit by exporting these technologies—if additional incentives are at all necessary for investing in the future of energy from the sun.

# Fusion

*Energy derived from fused nuclei may become widely used by the middle of the next century.*

• • •

Harold P. Furth

During the 1930s, when scientists began to realize that the sun and other stars are powered by nuclear fusion, their thoughts turned toward re-creating the process, at first in the laboratory and ultimately on an industrial scale. Because fusion can use atoms present in ordinary water as a fuel, harnessing the process could assure future generations of adequate electric power. By the middle of the next century, our grandchildren may be enjoying the fruits of that vision.

The sun uses its strong gravity to compress nuclei to high densities. In addition, temperatures in the sun are extremely high, so that the positively charged nuclei have enough energy to overcome their mutual electrical repulsion and draw near enough to fuse. Such resources are not readily available on the earth. The particles that fuse most easily are the nuclei of deuterium (D, a hydrogen isotope carrying an extra neutron) and tritium (T, an isotope with two extra neutrons). Yet to fuse even D and T, scientists have to heat the hydrogen gases intensely and also confine them long enough that the particle density multiplied by the confinement time exceeds $10^{14}$ seconds per cubic centimeter. Fusion research since the 1950s has focused on two ways of achieving this number: inertial confinement and magnetic confinement.

The first strategy, inertial confinement, is to shine a symmetrical array of powerful laser beams onto a spherical capsule containing a D-T mixture (see Figure 16.1). The radiation vaporizes the surface coating of the pellet, which explodes outward. To conserve momentum, the inner sphere of fuel simultaneously shoots inward. Although the fuel is compressed for only a brief moment—less than $10^{-10}$ second—extremely high densities of almost $10^{25}$

**Figure 16.1 LASER FUSION, here demonstrated at the Omega facility at the University of Rochester, is achieved by compressing a fuel pellet by means of symmetrically arranged lasers. The technology also has applications in defense.**

particles per cubic centimeter have been achieved at the Nova laser facility at Lawrence Livermore National Laboratory.

For more energetic lasers, the compression will be higher, and more fuel will burn. A future machine, the National Ignition Facility, the design and funding for which will be submitted to the U.S. Congress for final approval in 1996, will feature a 192-beam laser applying 1,800 kilojoules of energy within a few billionths of a second. If all goes well, fusion with this machine will liberate more energy than is used to initiate the capsule's implosion. France is planning to build a laser of similar capabilities near Bordeaux in 1996.

## Magnetic Fields

The many magnetic fusion devices explored—among them stellarators, pinches and tokamaks—confine the hot ionized gas, or plasma, not by material walls but by magnetic fields. The most successful and highly developed of these devices is the tokamak, proposed in the early 1950s by Igor Y. Tamm and Andrei D. Sakharov of Moscow University. Electric current flows in coils that are arranged around a doughnut-shaped chamber. This current acts in concert with another one driven through the plasma to create a magnetic field spiraling around the torus. The charged nuclei and an accompanying swarm of electrons follow the magnetic-field lines. The device can confine the plasma at densities of about $10^{14}$ fuel particles per cubic centimeter for roughly a second.

But the gas also needs to be heated if it is to fuse. Some heat comes from the electrical resistance to the current flowing through the plasma. But more intense heating is required. One scheme being explored in tokamaks around the world uses radiofrequency systems similar to those in microwave ovens. Another common tack is to inject energetic beams of deuterium or tritium nuclei into the plasma. The beams help to keep the nuclei hotter than the electrons. Because it is the nuclei that fuse, the available heat is therefore used more efficiently. (The strategy represents a departure from earlier experiments that tried to imitate the sun in keeping the temperatures of the nuclei and the electrons roughly equal.)

Such a "hot ion" mode was used in 1994 at Princeton University's Tokamak Fusion Test Reactor (TFTR) to generate more than 10 million watts of fusion power (see Figure 16.2). Although achieved for only half a second, the temperature, pressure and energy densities obtained are comparable with those needed for a commercial electrical plant. In 1996, during the next operational phase of the Joint European Torus (JET) at Culham, England, the experimenters may approach breakeven, generating as much energy as is fed into the plasma (see Figure 16.3). At the JT-60U device at Naka, Japan, scientists are developing higher-energy injectors.

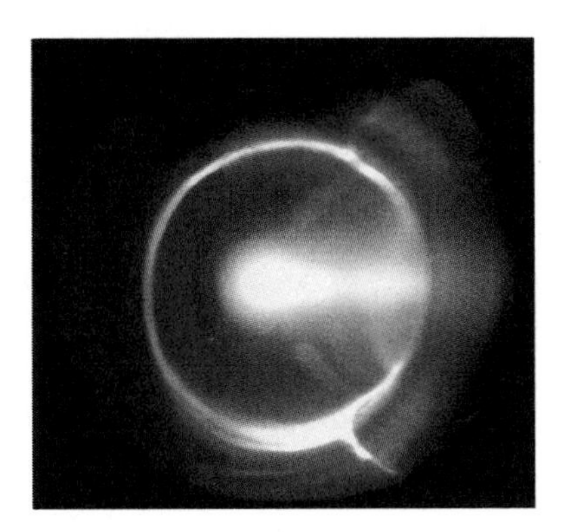

**Figure 16.2 TOKAMAK Fusion Test Reactor, a magnetic fusion machine at Princeton, N.J., has achieved the highest energies to date. The hot fuel, consisting of deuterium and tritium nuclei, is confined by magnetic-field lines. These lines are generated by electric currents flowing around a doughnut-shaped container. The hot gas causes the inner walls, made of carbon, to glow pink. The white streak is from deuterium being injected into the chamber.**

## Disposing of Nuclear Waste

At 3:49 P.M.on December 2, 1942, in a converted squash court under the football stands at the University of Chicago, a physicist slid back some control rods in the first nuclear reactor and ushered in a new age. Four and a half minutes later the world had its first nuclear waste. Since then, mountains of high-level waste have joined that original molehill in Chicago: according to the International Atomic Energy Agency, about 10,000 cubic meters of high-level waste accumulate each year.

This massive amount of radioactive material has no permanent home. Not a single country has managed to implement a long-term plan for storing it; each relies on interim measures. In the U.S., for example, used fuel rods are generally kept in pools near a reactor until they are cool enough for dry storage in steel casks elsewhere at the site.

However large or small a role nuclear fission plays in meeting future energy needs, safely disposing of intensely radioactive by-products will remain a top priority, if only because so much material already exists. Among the most promising technologies are:

• **Permanent subterranean storage.** All the countries that have significant amounts of high-level waste are currently hoping to store it deep underground in geologically stable areas. In the U.S. version of this plan, spent fuel rods are to be sealed in steel canisters and allowed to cool aboveground for several years. If the high-level waste is in liquid form, it will be dried and "vitrified," or enclosed in glass logs, before being put in canisters. These containers will later be placed in canisters up to 5.6 meters long, which will, in turn, be inserted in holes drilled in the rock floor of warehouselike caverns, hundreds of meters below the surface. The holes will be covered with plugs designed to shield the room above from radiation. Federal officials had hoped to build the U.S. repository beneath Yucca Mountain in Nevada, but local authorities are fighting to keep the repository out of the state.

• **Entombment under the seabed.** Pointed canisters containing the waste could be dropped from ships to the floor far below, where they would penetrate and embed themselves tens of meters down. The advantages stem from the ability to use seafloor sites that are stable and remote from the continents. In a variation on this idea, the canisters could be dropped into deep ocean trenches, where they would be pulled into the earth's mantle by the geologic process of subduction.

Considered the most scientifically sound proposals by some experts, these—as with all schemes involving the oceans—have been rejected by science policymakers because of concerns over the potentially adverse public reaction and the plan's possible violation of international treaties barring the disposal of radioactive waste at sea.

• **Nuclear transmutation.** The troublesome components of high-level waste are a relatively small number of materials that are radioactive

Apart from keeping the plasma hot, one challenge is to clean out continuously the contaminant atoms that are knocked off when nuclei happen to strike the walls. Several tokamaks have extra magnetic coils that allow the outer edges of the plasma to be diverted into a chamber where the impurities are extracted along with some heat. This system works well for present-day experiments, in which the plasma is confined for at most a few seconds. But it will not suffice for commercial power plants that will generate billions of watts during pulses that last hours or days. Researchers at JET and at the DIII-D tokamak in San Diego are attacking this problem.

particles per cubic centimeter have been achieved at the Nova laser facility at Lawrence Livermore National Laboratory.

For more energetic lasers, the compression will be higher, and more fuel will burn. A future machine, the National Ignition Facility, the design and funding for which will be submitted to the U.S. Congress for final approval in 1996, will feature a 192-beam laser applying 1,800 kilojoules of energy within a few billionths of a second. If all goes well, fusion with this machine will liberate more energy than is used to initiate the capsule's implosion. France is planning to build a laser of similar capabilities near Bordeaux in 1996.

## Magnetic Fields

The many magnetic fusion devices explored—among them stellarators, pinches and tokamaks—confine the hot ionized gas, or plasma, not by material walls but by magnetic fields. The most successful and highly developed of these devices is the tokamak, proposed in the early 1950s by Igor Y. Tamm and Andrei D. Sakharov of Moscow University. Electric current flows in coils that are arranged around a doughnut-shaped chamber. This current acts in concert with another one driven through the plasma to create a magnetic field spiraling around the torus. The charged nuclei and an accompanying swarm of electrons follow the magnetic-field lines. The device can confine the plasma at densities of about $10^{14}$ fuel particles per cubic centimeter for roughly a second.

But the gas also needs to be heated if it is to fuse. Some heat comes from the electrical resistance to the current flowing through the plasma. But more intense heating is required. One scheme being explored in tokamaks around the world uses radio-frequency systems similar to those in microwave ovens. Another common tack is to inject energetic beams of deuterium or tritium nuclei into the plasma. The beams help to keep the nuclei hotter than the electrons. Because it is the nuclei that fuse, the available heat is therefore used more efficiently. (The strategy represents a departure from earlier experiments that tried to imitate the sun in keeping the temperatures of the nuclei and the electrons roughly equal.)

Such a "hot ion" mode was used in 1994 at Princeton University's Tokamak Fusion Test Reactor (TFTR) to generate more than 10 million watts of fusion power (see Figure 16.2). Although achieved for only half a second, the temperature, pressure and energy densities obtained are comparable with those needed for a commercial electrical plant. In 1996, during the next operational phase of the Joint European Torus (JET) at Culham, England, the experimenters may approach breakeven, generating as much energy as is fed into the plasma (see Figure 16.3). At the JT-60U device at Naka, Japan, scientists are developing higher-energy injectors.

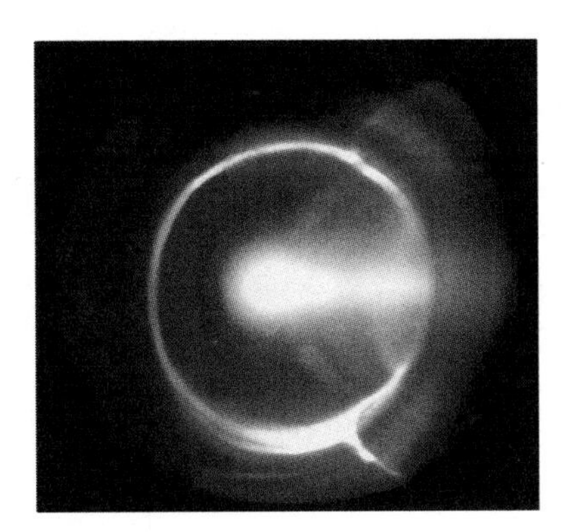

**Figure 16.2 TOKAMAK Fusion Test Reactor, a magnetic fusion machine at Princeton, N.J., has achieved the highest energies to date. The hot fuel, consisting of deuterium and tritium nuclei, is confined by magnetic-field lines. These lines are generated by electric currents flowing around a doughnut-shaped container. The hot gas causes the inner walls, made of carbon, to glow pink. The white streak is from deuterium being injected into the chamber.**

## Disposing of Nuclear Waste

At 3:49 P.M.on December 2, 1942, in a converted squash court under the football stands at the University of Chicago, a physicist slid back some control rods in the first nuclear reactor and ushered in a new age. Four and a half minutes later the world had its first nuclear waste. Since then, mountains of high-level waste have joined that original molehill in Chicago: according to the International Atomic Energy Agency, about 10,000 cubic meters of high-level waste accumulate each year.

This massive amount of radioactive material has no permanent home. Not a single country has managed to implement a long-term plan for storing it; each relies on interim measures. In the U.S., for example, used fuel rods are generally kept in pools near a reactor until they are cool enough for dry storage in steel casks elsewhere at the site.

However large or small a role nuclear fission plays in meeting future energy needs, safely disposing of intensely radioactive by-products will remain a top priority, if only because so much material already exists. Among the most promising technologies are:

• **Permanent subterranean storage.** All the countries that have significant amounts of high-level waste are currently hoping to store it deep underground in geologically stable areas. In the U.S. version of this plan, spent fuel rods are to be sealed in steel canisters and allowed to cool aboveground for several years. If the high-level waste is in liquid form, it will be dried and "vitrified," or enclosed in glass logs, before being put in canisters. These containers will later be placed in canisters up to 5.6 meters long, which will, in turn, be inserted in holes drilled in the rock floor of warehouselike caverns, hundreds of meters below the surface. The holes will be covered with plugs designed to shield the room above from radiation. Federal officials had hoped to build the U.S. repository beneath Yucca Mountain in Nevada, but local authorities are fighting to keep the repository out of the state.

• **Entombment under the seabed.** Pointed canisters containing the waste could be dropped from ships to the floor far below, where they would penetrate and embed themselves tens of meters down. The advantages stem from the ability to use seafloor sites that are stable and remote from the continents. In a variation on this idea, the canisters could be dropped into deep ocean trenches, where they would be pulled into the earth's mantle by the geologic process of subduction.

Considered the most scientifically sound proposals by some experts, these—as with all schemes involving the oceans—have been rejected by science policymakers because of concerns over the potentially adverse public reaction and the plan's possible violation of international treaties barring the disposal of radioactive waste at sea.

• **Nuclear transmutation.** The troublesome components of high-level waste are a relatively small number of materials that are radioactive

Apart from keeping the plasma hot, one challenge is to clean out continuously the contaminant atoms that are knocked off when nuclei happen to strike the walls. Several tokamaks have extra magnetic coils that allow the outer edges of the plasma to be diverted into a chamber where the impurities are extracted along with some heat. This system works well for present-day experiments, in which the plasma is confined for at most a few seconds. But it will not suffice for commercial power plants that will generate billions of watts during pulses that last hours or days. Researchers at JET and at the DIII-D tokamak in San Diego are attacking this problem.

**SPENT FUEL RODS are cooled and shielded in pools at a facility in La Hague, France. Blue glow is caused by the interaction of radiation from the fuel with the water.**

for tens of thousands of years. If appropriately bombarded with neutrons, however, the materials can be transmuted into others that are radioactive for only hundreds or possibly just tens of years. A repository would still be needed. It could hold much more material, though, because the amount of heat emitted by the waste would be significantly reduced.

A very small scale form of transmutation has been carried out for decades in specially designed experimental reactors. Lately scientists at Los Alamos National Laboratory have proposed using a high-energy accelerator to make the process more rapid and efficient. According to Wendell D. Weart, a senior scientist at Sandia National Laboratories, the main challenge would be concentrating the nuclear materials. "No one's ever tried to do it with the degree of separation that this would require," he says. "It's not easy to work with intensely radioactive systems, with this degree of chemical separation. It would be a neat trick."

Some less highly regarded proposals are:

• **Shooting nuclear waste into space or the sun.** This idea has been rejected because of its enormous expense, as well as the possibility that a loaded rocket could blow up before leaving the earth's atmosphere.

• **Storage under a polar icecap.** High-level waste generates enough heat to melt not only ice but possibly even rock. Perhaps because of that characteristic, the idea has not won many converts.

• **Dissolving waste in the world's oceans.** Uniformly spread over much of the surface of the earth, the radioactivity would be small compared with the background level, proponents have insisted.

*—Glenn Zorpette, Scientific American*

Currently it is within our capability to construct and operate a tokamak that will sustain a stable, fusing plasma, not for fractions of a second but for thousands of seconds. The International Thermonuclear Experimental Reactor (ITER), a collaborative effort of the European Union, Japan, the Russian Federation and the U.S., aspires to do just that. ITER is expected to be a large machine, with a plasma about 16 meters in major diameter, featuring superconducting coils, tritium-breeding facilities and remote maintenance. The present schedule calls for a blueprint to be completed in 1998, whereupon the participating governments will decide whether to proceed with construction.

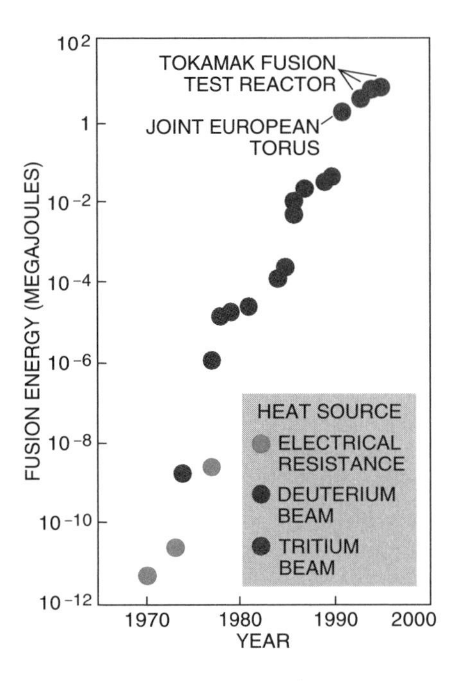


**Figure 16.3 ENERGY RELEASE from fusion in tokamaks around the world has increased steadily in the past decades. Much of the improvement has come from heating the fuel by injection of energetic nuclei.**

## Looking Ahead

While ITER represents a valuable opportunity for international collaboration, there is some chance that its construction will be delayed. In the interim, experimenters at TFTR and other large tokamaks around the world will explore new ideas for making radically smaller and cheaper reactors, possibly influencing ITER's design. One of these proposals uses a complex twisting of the plasma to reduce heat loss greatly. Normally the magnetic field that winds around the torus has a higher twist near the center of the plasma. If instead the twist decreases in regions near the center, the plasma should be less turbulent, thus allowing higher pressures to be sustained. Experimentation, along with theoretical analysis and computer simulations, will vastly improve our ability to control such processes.

One more hint of how fusion could evolve in the next 25 years is in the utilization of the by-products. In a self-sustaining reaction the heat lost is made up by energy generated through fusion. Eighty percent of the fusion power is carried away by neutrons, which, being electrically neutral, slip through the confining magnetic fields. Trapped in outer walls, the neutrons give up their energy as heat, which is used to generate steam. The vapor in turn drives an engine to create electricity. The remaining 20 percent of the power goes to the other product of the fusion reaction, an alpha particle: a pair of protons and a pair of neutrons, bound into a nucleus. Being positively charged, alphas are trapped by the magnetic fields.

The alphas bounce around inside the plasma, heating up the electrons; the fuel nuclei are heated indirectly through collisions with electrons. Nathaniel J. Fisch of Princeton and Jean-Marcel Rax of the University of Paris suggested in 1992 that rather than wasting valuable energy on the electrons, the alpha particles could help amplify special waves injected into the plasma that channel the energy directly to the nuclei. Thus concentrating the energy in the fuel could double the density of fusion power achieved.

The particles produced as by-products of fusion may be put to another, quite different, use. In this respect, taking a hint from history might be beneficial for fusion's short-term future. Two centuries ago in England, the industrial revolution came about because horses refused to enter coal mines: the first engines were put together to haul out coal, not to power cars or airplanes. John M. Dawson of the University of California at Los Angeles has proposed that during the next 20 to 30 years, while fusion programs are developing a technology for large-scale energy production, they could provide other benefits. For example, the protons formed as by-products of some fusion reactions may be converted to positrons, particles that can be used in medically valuable positron emission tomography scans.

During this phase of special applications, an abundance of new ideas in plasma physics would be explored, ultimately yielding a clear vision about future reactor design. Fifty years from now engineers should be able to construct the first industrial plants for fusion energy. Although far removed from immediate political realities, this schedule matches the critical timescale of 50 to 100 years in which fossil-energy resources will need to be replaced.

# The Industrial Ecology of the 21st Century

*A clean and efficient industrial economy would mimic the natural world's ability to recycle materials and minimize waste.*

• • •

Robert A. Frosch

The end of the 20th century has seen a subtle change in the way many industries are confronting environmental concerns: they are shifting away from the treatment or disposal of industrial waste and toward the elimination of its very creation. This strategy attempts to get ahead of the problem, so that society is not destined to face an ever growing mass of waste emanating from the end of a discharge pipe or the brim of a garbage pail. It seems likely that the next century will see an acceleration of this trend, a clear departure from the past emphasis—by industry, by government regulators and even by most environmental organizations—on late-stage cleanup.

The old attitude often resulted only in manufacturers dumping waste into their own "backyards," thus generating a good deal of what might be called industrial archaeology. That heritage currently puts many firms into the environmental cleanup business, whether they like it or not. But in the 21st century, industry may behave quite differently, so as to avoid creating more expensive burial sites that society will have to suffer or pay to clean up all over again.

What most people would like to see is a way to use industrial waste productively. Waste is, after all, *wasteful.* It is money going out the door in the form of processed material and its embodied energy. To avoid this inefficiency, manufacturers of the next century must consider how to design and produce products in such a way as to make the control of waste and pollution part of their enterprise, not just an afterthought. They will need to pay attention to the entire product life cycle, worrying not only about the materials used and created in the course of manufacturing but also about what happens to a product at the end of its life. Will it become a disposal problem, or can it become a source of refined material and energy?

Manufacturers are just beginning to seek new approaches in what may well become a comprehensive revolution. As such movements often do, these efforts are producing new ideas and a new set of buzzwords. Engineers had previously spoken of "design for manufacturing" and "design for assembly," and now we have added "design for disassembly," "design for recycling" and "design for environment" to our vocabulary. These terms mean simply that from the very start we are paying attention to the potential effects of excess waste and pollution in manufacturing.

Overcoming these problems is in part a technological problem—clever new technologies that can reduce or recycle wastes will surely play a valuable role. But the answer will not depend entirely on inventing breakthrough technologies. Rather it may hinge on coordinating what are fairly conventional methods in more prudent ways and in developing legal and market structures that will allow suitable innovation. These efforts will involve complex considerations of product and process design, economics and optimization, as well as regulation and handling of hazardous materials. Strangely, there has been relatively little general examination of these issues, although there are many individual cases in which such thinking has been employed.

For example, Kumar Patel of AT&T Bell Laboratories has described an interesting approach being taken in a section of its microelectronics fabrication business. Engineers at that division of Bell Labs were concerned because several of the raw materials, such as gallium arsenide, were particularly nasty. They dealt with this difficulty by using, in effect, the military technology of binary chemical weapons, in which two chemicals that are not very hazardous individually combine within a weapon to make one tremendously hazardous substance. Bell Labs now avoids having to keep an inventory of one highly toxic material through a simple process that brings together its much less hazardous chemical constituents right at the spot where the combined compound is used. This is essentially a just-in-time delivery system that matches production to need and obviates disposal of excess. Bell Labs concluded that the company amortized the investment in new equipment in less than a year by eliminating the extra costs of storing, transporting and occasionally disposing of the hazardous compound.

## A Lesson from Nature

Beyond solving as much of the waste problem as possible within each company, we have to think about industry in the future on a larger scale. We need to examine how the total industrial economy generates waste and pollutants that might damage the environment. Viewing industry as an interwoven system of production and consumption, one finds that the natural world can teach us quite a bit. The analogy with nature suggests the name "industrial ecology" for this idea (although this term is increasingly coming into use for a diverse set of practices that might make industry pollute less).

The natural ecological system, as an integrated whole, minimizes waste. Nothing, or almost nothing, that is produced by one organism as waste is not for another organism a source of usable material and energy. Dead or alive, all plants and animals and their wastes are food for something. Microbes consume and decompose waste, and these microorganisms in turn are eaten by other creatures in the food web. In this marvelous natural system, matter and energy go around and around in large cycles, passing through a series of interacting organisms.

With this insight from the natural ecological system, we are beginning to think about whether there are ways to connect different industrial processes that produce waste, particularly hazardous waste. A fully developed industrial ecology might not necessarily minimize the waste from any specific factory or industrial sector but should act to minimize the waste produced overall. For example, Kalundborg, Denmark, 60 miles west of Copenhagen, represents a model industrial ecosystem. Its oil refinery employs waste heat from a power plant and sells sulfur removed from petroleum to a chemical company. The refinery also provides sulfur (as calcium sulfate) to a wallboard producer to replace the gypsum typically used. Excess steam from the power plant also heats water for aquaculture while it warms greenhouses and residences.

This is not really a new or startling idea. There are companies that have sought this minimization for a long time (see Figure 17.1). The chemical and petrochemical industries are probably ahead of most others. They characteristically think in terms of turning as much as possible of what they process into useful product. But in the future, industrial countries will want all producers to be thinking about how they can alter manufacturing, products and materials so that the ensemble minimizes both waste and cost. Such requirements need not be onerous: a company might easily change to a more expensive manufacturing process if it prevents the generation of waste that the firm had to pay to have taken away and if it creates materials for which there are customers.

Many requirements must be met for this redirection to be accomplished. As incentive for designing and producing something specifically so that it can be reused, companies will need reliable markets

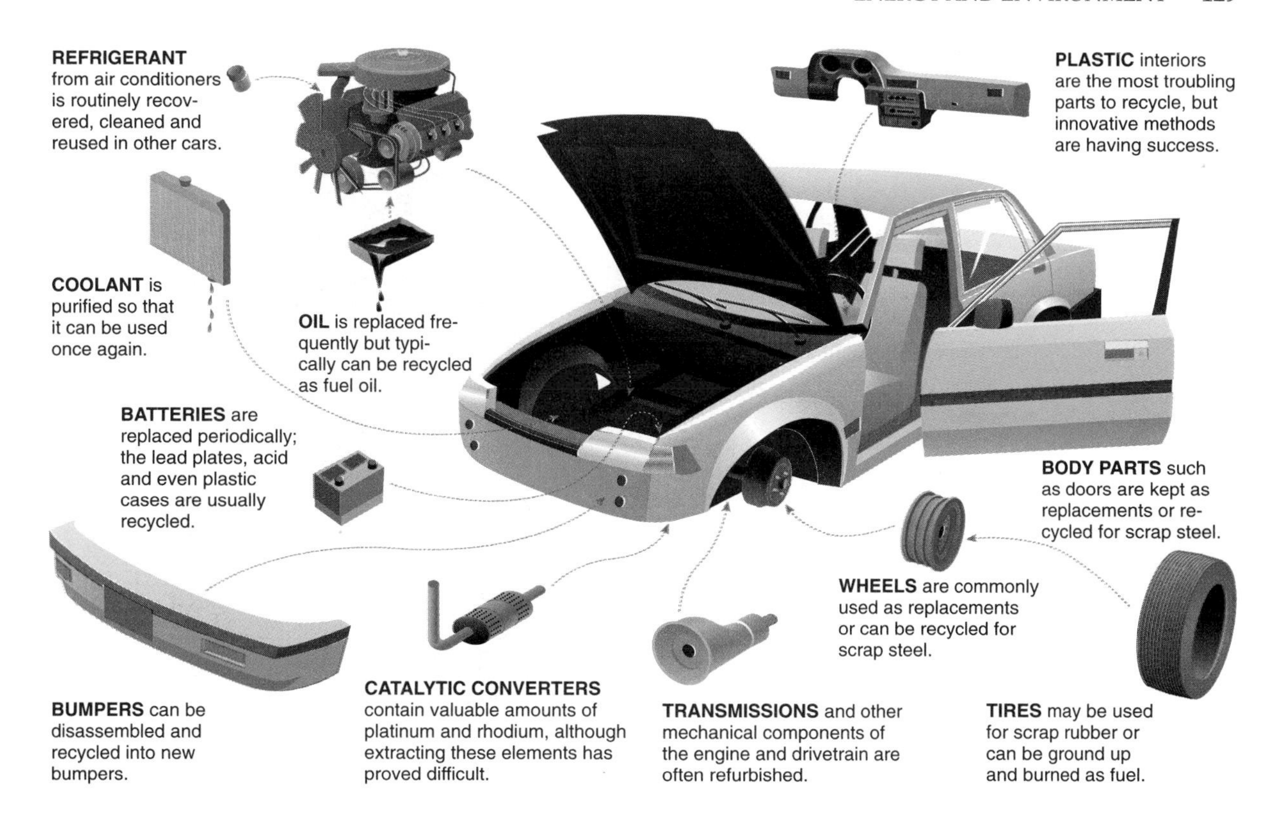


**Figure 17.1 AUTOMOBILE RECYCLING is one of the most successful examples of reuse of a manufactured product. About 75 percent of a typical car can be recycled in the form of refurbished parts, useful fluids and scrap materials. This process can, however, be taken further. New methods to separate and recycle plastic components, for example, offer the possibility of removing even more material from the waste stream and returning it to the manufacturing cycle.**

Many early attempts at recycling failed because they just collected materials—a pointless exercise unless somebody actually wants to use them. If there are going to be markets for what would otherwise be waste, information will need to be available on who has what, who needs what, who uses what. This information is typically inaccessible now because companies tend to be secretive about their waste streams. (If competitors know about the by-products produced, they might deduce protected trade secrets.) We will have to invent ways to get around this difficulty.

## Antirecycling Laws

In addition to the need for more complete market information, society requires a novel kind of regulation to make a true industrial ecology possible. Frustrations with regulation frequently arise because we have fostered and developed environmental laws that attempt to deal with one problem at a time. The current regulatory framework focuses on disposing of or treating industrial wastes without regard for the possibility of minimizing or reusing them. In fact, it often acts to thwart recycling. Once a substance is classified as hazardous waste, it becomes extraordinarily difficult to do anything useful with it, even if the material is identical to a "virgin" industrial chemical readily bought and sold on the open market.

For example, if a manufacturer produces waste containing cyanide, a toxic hydrocarbon or a heavy metal, the company will likely be controlled by strict environmental laws. Unless the firm can overcome excruciatingly complex bureaucratic barriers, it will probably not be allowed to process that material into a salable product or even to transport it (except to a disposal site). Yet anyone can easily go to a chemical manufacturer and buy cyanide,

## The Ultimate Incinerators

While the world waits for industry to develop processes so efficient they do not produce waste, the problem of safely disposing of our garbage persists. The idea of loading toxic or other forms of waste on board a spacecraft and blasting them into the sun seems, at first glance, a nice solution to the earth's trash woes. At 5,500 degrees Celsius, the surface of Sol would leave little intact. But considering the amount of garbage each human produces—three to four pounds per day, on average—launches would simply be too expensive to conduct regularly. Add the possibility of a malfunction during liftoff, and space shots of waste seem impractical.

Instead some researchers are taking the opposite tack: bringing a bit of the sun to the earth. By sending a strong electric current through a rarefied gas, they can create plasma—an intensely hot gas in which electrons have been separated from atomic nuclei. The plasma, in turn, reaches up to 10,000 degrees C. (Conventional incinerators, using fossil fuels, reach no more than 2,000 degrees C.) In the presence of this demonic heat, hydrocarbons, PCBs and other toxins that lace contaminated soil and ash break down, yielding molten slag that hardens into inert and harmless glassy rocks suitable for road gravel. Unlike their smoke-belching conventional counterparts, plasma incinerators burn more cleanly, emitting one fifth as much gas. Some designers propose capturing this gas, which is combustible, for use as fuel.

**PLASMA TORCH cooks contaminated soil, changing it into inert, glassy blocks.**

With so many pluses, it seems that plasma should have been cooking waste a long time ago.

hydrocarbon solvents or heavy metal compounds that have been newly produced. (Their manufacturer generally has a standing permit for packaging, transporting and selling these substances.)

A particularly interesting example comes from the automotive industry's treatment of steel. Anticorrosion measures produce a zinc-rich sludge that in the past was sent to a smelter to recover the zinc and put it back into the process stream. But a decade ago regulations began listing such wastewater treatment sludges as hazardous. The unintended consequence was that the smelters could no longer use the sludge, because it had become, in name, a hazardous material—the regulatory requirements for accepting it were too severe. The zinc-rich sludge was redirected to landfills, thereby increasing costs for automobile manufacturers and producing a waste disposal problem for the rest of society.

This situation clearly illustrated what can be a serious problem: well-meant environmental regulation can have the bizarre effect of increasing both the amount of waste created and the amount to be

The hurdle has been economic: plasma can vaporize nonhazardous waste for about $65 a ton, whereas landfilling costs less than half that amount. But as landfill space dwindles and stricter environmental codes are adopted, plasma waste destruction is becoming more competitive. For treatment of toxic waste, it may even be cheaper. Daniel R. Cohn of the Massachusetts Institute of Technology estimates that a full-scale plant could operate for less than $300 a ton—less than half the current cost of disposing of hazardous waste.

The more reasonable economics have encouraged many institutions to set up pilot furnaces. "The whole technology is starting to pick up around the world," notes Louis J. Circeo of the Georgia Institute of Technology, where some of the largest furnaces are located. Near Bordeaux, France, a plant destroys asbestos at the rate of 100 tons a week. The Japanese city of Matsuyama has a facility designed to handle the 300 tons of incinerator ash that comes from the daily burning of 3,000 tons of municipal waste. Construction of a furnace that could torch 12 tons of medical waste a day is under way at Kaiser Permanente's San Diego hospital. Circeo thinks it is even feasible to treat existing landfills: just lower some plasma torches down nearby boreholes.

Plasma need not be hot; it can also exist at room temperature. Cohn and his colleagues are testing the idea of using "cold" plasma to destroy toxic vapors. The physicists create such plasma by firing an electron beam into a gas, a process that severs electrons from nuclei and thus converts the gas into plasma. Volatile organic compounds passed through the plasma are attacked by the free electrons, which break down the chemicals. Last year the workers tested their trailer-size unit at the Department of Energy's Hanford Nuclear Reservation site in Richland, Wash., where up to two million pounds of industrial solvents have been dumped since the complex's founding during the Manhattan Project. They vacuumed out some of the carbon tetrachloride in the ground and then pumped it into the chamber of cold plasma, which transformed the toxin into less harmful products that were subsequently broken down into carbon dioxide, carbon monoxide, water and salt.

It may be a while before toxic waste is a distant memory or before you can zap your kitchen trash into nothingness with the flick of a switch, but many researchers are betting that plasma waste destruction is becoming a reality. Circeo, for instance, hopes to raise $10 million for a plasma plant that can destroy the 20 tons of garbage that revelers and others at the 1996 Olympic Games in Atlanta are expected to generate daily. "In five to 10 years," he predicts, "you're going to see plasma technology springing up all over the place."

—*Phil Yam,* SCIENTIFIC AMERICAN

disposed, because it puts up high barriers to reuse. It might be viewed as antirecycling regulation. This peculiarity appears to have occurred essentially by inadvertence: industrial supplies, whether toxic or not, are controlled by different statutes—and often by a different part of the government—than are materials considered waste. A priority for the future will be a cleanup of that aspect of the nation's regulatory machinery.

With adequate effort the next century will see many improvements in environmental laws as well as in specific environmental technologies. But the most important advance of all may be the fundamental reorganization that allows used materials to flow freely between consumers and manufacturers, between one firm and the next and between one industry and another. As much as we need to excavate the industrial archaeology left over from the past, we also need to draw lessons for the future from these ghastly sites, create an industrial ecological vision and formulate a system of law and practice to enable it.

# Technology for Sustainable Agriculture

*The next green revolution needs to be sophisticated enough to increase yields while also protecting the environment.*

• • •

Donald L. Plucknett and Donald L. Winkelmann

Farmers will have to confront formidable challenges in learning to manage ever more advanced technologies in ways that will increase the productivity of their resources while protecting the environment. That complex goal should be within reach of richer countries, where (ironically) food is already abundant and affordable, population growth is slow and mechanisms exist for resolving at least some environmental problems associated with agriculture. But it will surely be a daunting task in the developing world, where about a billion people are being added each decade, where roughly that number are already malnourished and where social capital for environmental protection is severely limited.

Many of the earth's less developed regions are indeed in critical condition. In the poorest countries, agriculture now occupies up to 80 percent of the workforce, and almost half of an average family's budget is spent on foodstuffs. In those nations, incomes can grow only with greater farm productivity. Advances in agriculture thus become a prerequisite for achieving nearly all social goals—including slower population growth.

Technology has been the most reliable force for pushing agriculture toward higher productivity throughout this century. The many agricultural technologies in use today all emerged from scientific research. So it is reasonable to ask what science will offer the next generation of farmers and consumers. How, in particular, might research in the next few decades help in the poorest countries, where poverty frequently causes farmers to jeopardize the health of the environment? The answer, simply put, is by fostering the practice of sustainable agriculture.

For our purposes, we define sustainable agriculture as does Pierre Crosson of Resources for the Future: it is farming that meets rising demands over the indefinite future at economic, environmental and other social costs consistent with rising incomes. Sustainable agriculture is thus quite a broad topic; we concentrate here on just a few of its more technological aspects.

## Damage Control

Wherever agriculture is practiced, insects and diseases lurk and threaten crops. Every cultivated field in the world also harbors weeds, and a huge amount of labor, often by women and children, is required to control them. Since the 1950s, the treatment for crops suffering from such pests has sometimes been an indiscriminate application of potent chemicals. But concern for environmental standards in richer countries is beginning to decrease their use. Even some safe, commonly used pesticides will soon become unavailable because, when their patents run out, the cost of complying with environmental regulations in marketing these products may exceed the profits that might otherwise accrue to their manufacturers. Fortunately, these chemicals are not the only weapons available to the battle. A wide array of alternative measures can be brought in for support. The term "integrated pest management" captures that notion. It refers to the combination of using hardy plants, crop rotations, tillage practices, biological controls and a minimal amount of pesticides.

Integrated pest management draws on the fundamental knowledge of plant and insect biology amassed by botanists and entomologists. Hundreds of insect attractants (pheromones) have already been identified and synthesized, and these substances can be used to interfere with the normal reproductive cycle of common pests—for example, by inducing the insects to mate too soon or inappropriately. Researchers are also enthusiastic about the possibility of using insect viruses to suppress pests effectively without harming their natural predators or leaving unwanted chemical residues. Coming advances will undoubtedly make such carefully tailored insect-control techniques more widely available to knowledgeable farmers, who can then incorporate them as standard routines.

As scientific knowledge grows in the next century, researchers will continue to turn out a host of clever inventions that can challenge conventional pesticides. But the success of alternative measures will also require a clear understanding of farm practices. For example, it might be possible, in principle, to reduce insect, disease and weed problems by increasing tillage or removing crop residues from some fields. But such actions can also increase erosion. Hence, integrated pest management will increasingly test the skills of both researchers and farmers to anticipate such effects and to balance what are sometimes competing goals. Despite the increased care in management that will be required of farmers, we are nonetheless optimistic that future efforts will provide a spectrum of practical strategies for effective pest control.

## Power Plants

The success of sustainable agriculture depends fundamentally on making plants more efficient in converting sunlight, nutrients and water into food and fiber products. Conventional plant breeding now boosts yields by roughly 1 percent annually. In the coming years, biotechnology can be expected to make breeding even more efficient. For example, research on the DNA of diverse organisms will reveal the genetic basis of many traits, including disease and insect susceptibility, biochemical composition and nutritive value. Breeders will be able to use genetic tests and markers to identify subtle but desirable traits in their crops more readily. Biotechnology is also opening the door to completely new hybrids by making it possible to crossbreed plants that might never be able to do so naturally. Moreover, the application of modern molecular biology will allow desirable traits to be directly "engineered" into crop lines.

We expect that genetic manipulation will augment conventional breeding by allowing at first the direct transfer of selected genes to existing food crops from their wild relatives. Through this approach, there will be continued progress in bioengineering various plants, for instance, to withstand diseases or pests (improving "host plant resistance"), as well as to resist stress from high-acidity soil, drought or toxic elements. We are optimistic about seeing these techniques increase the yield of food crops; in particular, they may help boost productivity on some of the developing world's marginal lands.

Efforts in promoting host plant resistance will benefit also from new insights into the molecular biology of viruses. Scientists have already inserted genes from plant viruses into the genetic makeup of selected crops, creating inherent resistance to these viral diseases in the altered plants. During the next century, such techniques—ones that bring exotic genes into common crops—will be extensively developed.

A forerunner of the marvels that the future holds is the incorporation of a gene from the bacterium

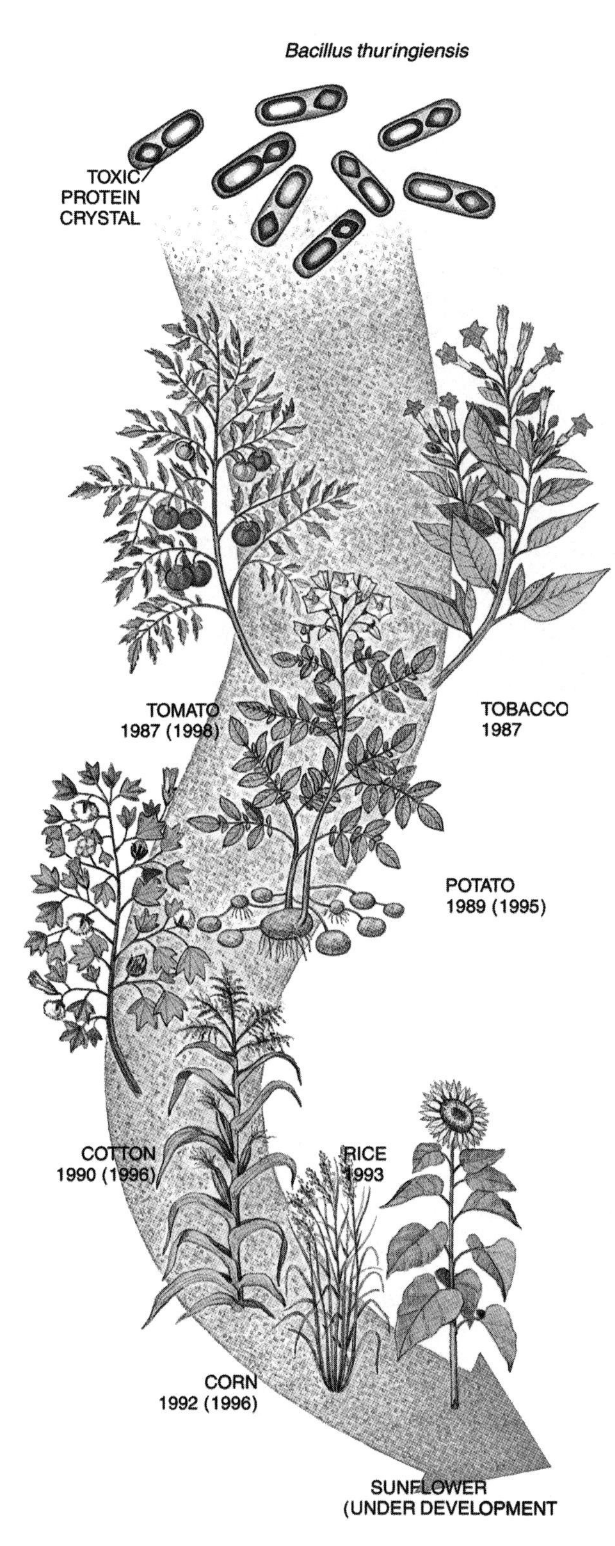


*Bacillus thuringiensis* into a variety of crops (see Figure 18.1). The foreign gene codes for a toxin that gives plants resistance to some leaf-feeding caterpillars. The transfer of traits across species lines is a premier example of the way technology can improve the yield of various crops without adversely altering or degrading the growing environment.

We expect that during the next few decades genetic engineering will become a powerful addition to proved plant-breeding strategies. But if the researchers of the next century are going to be able to mix and match genes to produce better crops, they are going to need a wide variety of natural material to work with.

## Applied Biodiversity

The conservation of biodiversity is receiving ever more attention, in part because it affects agriculture so directly. For most economically important crops, researchers have carefully maintained collections of genetic material from domesticated and wild species in germ plasm banks. These repositories provide good insurance that the appropriate genes can be found and introduced into cultivated plants when new challenges arise.

Although a majority of the crops of economic value can be conserved by keeping their dry seeds refrigerated at about –10 degrees Celsius, about 10 percent of the plants now used for agriculture, along with a good proportion of the wild relatives of such crops, cannot withstand these storage conditions. Preservation of those varieties demands that the entire plant be maintained in something like its natural environment. Such "in situ" conservation may also be required where the number of species of interest is large—for example, in preserving the plants that inhabit grasslands or tropical forests. This strategy also maintains species that play symbiotic or associative roles—interactions that may prove economically important as well.

**Figure 18.1** ***BACILLUS THURINGIENSIS*, a rod-shaped bacterium, produces a protein that is toxic to leaf-eating caterpillars and so has been used for decades by organic farmers as a natural insecticide. The gene that codes for the toxin has been genetically engineered into several crops over the past decade and will undoubtedly be used in many more in the future. (Dates of experimental demonstration are shown beside projected dates for commercialization, which are given in parentheses.)**

## The Next Wave: Aquaculture

**AQUACULTURE FARMS can be found all over the world. Here, off the coast of Monaco, farmers are raising marine species in floating pens.**

Global consumption of fish is at its highest recorded level and is still growing. All forms of water dwellers, from fin fish to crustaceans, are in great demand. At the same time, however, scientists and fisheries experts are noticing a slow but worrisome decline in piscine populations: it appears the oceans are in danger of depletion because of poor management practices established decades ago and technologies that permit gargantuan catches.

Aquaculture—the cultivation of certain aquatic animals and plants in farms on land and at sea—may offer, at the very least, a partial solution to the problem of shrinking supply. The

Farmers in the next century, perhaps above all others, will welcome biodiversity because its practical applications may be as close as the next growing season. But if they are to utilize newly created varieties to their fullest extent, they must also guard the health of the environment in which food crops grow.

### A Good Working Environment

In the past, half the dramatic improvements in yield have come through the use of more fertilizers, more pesticides and more irrigation. But further increasing the amounts of such additions is not a formula to follow indefinitely. Farmers in the 21st century will have to acquire a new mind-set to manage their land as well as the nutrients and organic matter it contains (see Figure 18.2). This precept is especially true for farmers in developing countries, for whom immediacy normally rules most decisions.

In developed countries with temperate climates, many agricultural researchers are cautiously optimistic that farmers can reduce their use of chemicals and lower costs by better managing the soil's fauna and flora. With proper information and care, they can hasten the accumulation of organic matter

Worldwatch Institute in Washington, D.C., reports that farms are currently the fastest-growing source of fish. In 1993 aquaculture supplied about 22 percent of the 86 million metric tons caught, according to a study by the World Bank. More than 25 percent of the salmon and shrimp that are consumed come from farms. These successes have already made aquaculture a $26-billion industry. The World Bank estimates that aquaculture could meet some 40 percent of the demand for fish by 2005 if the proper investments in research and technology are made by governments today.

Aquaculture's potential is not news to many outside the U.S., Canada and Europe. Fish farming, particularly for tilapia (a freshwater cichlid fish) and shrimp, is common in many countries. Among the top producers are China, Japan, Norway, Israel, India, Ecuador, Thailand and Taiwan. Many of these countries "are reflecting about 20 years of research and development and investment," notes Michael C. Rubino, a partner at a shrimp farm, Palmetto Aquaculture in South Carolina, and a consultant at the International Finance Corporation. "Species take about 20 years to get from the wild into the barnyard, so to speak." (American efforts have centered on catfish, largely in the South, as well as oysters and salmon.)

Those practicing aquaculture have to contend with many of the same problems that any farmer faces: disease, pollution and space. When thousands of animals—be they fish or fowl—cohabit in proximity, they become highly susceptible to bacteria and viruses, which can wipe out an entire farm. Further, if water supplies are not clean, the harvest can suffer. For instance, according to Robert Rosenberry of Shrimp News International, shrimp production in China has fallen about 70 percent because of industrial pollution and poor water management (a loss of about $1 billion). Marine farms can also pollute if there is not enough circulation: fish feces and food can accumulate on the bottom and wipe out the benthic communities needed to filter the water.

Pollution emanating from these sites and the movement of diseases across ecosystems have turned some environmentalists off aquaculture. In their view, contamination is not the only problem. Marine aquaculture has contributed to the destruction of coastal wetlands, and many are concerned that carefully bred farm fish may escape and mate with wild stocks—raising the possibility that they could threaten biodiversity by limiting the gene pool.

Nevertheless, these woes may be addressed by the growing recognition that aquaculture, like any form of agriculture, must become sustainable. The environmental concerns are also seemingly outweighed by the potential of aquaculture to alleviate the demand for ocean fish and the diminishing wild stock. In fact, Rosenberry says, in terms of shrimp, it has already taken pressure off.

*—Marguerite Holloway,*
*SCIENTIFIC AMERICAN*

and, perhaps, better coordinate nutrient cycles with their crops' needs. But farmers in developing countries with tropical climates face more acute problems in managing their natural resources. This challenge emerges because poverty itself limits farmers' concern for long-term consequences and also because warm temperatures and high rainfall make it inherently difficult to protect land and water. Such climates tend to foster diseases and pests that have traditionally been difficult to control without broadly active chemicals—agents that may ultimately compromise the environment.

Research can help by bringing a better understanding of the biology and ecology on tropical farms—the soil biota, plant nutrient requirements, interactions with insects and pathogens—and these results should suggest more sustainable practices. In time, the range of advancement possible through better science could be enormous. In many cases, however, the people farming the land may be too poor to accept any new method very readily.

Societies concerned with protecting the environment need to find ways to raise the incentives for farmers to turn to measures that might conserve resources. Better financial and bureaucratic support, for instance, might give farmers the impetus they need to adopt sustainable methods. Such changes would greatly improve the outlook for

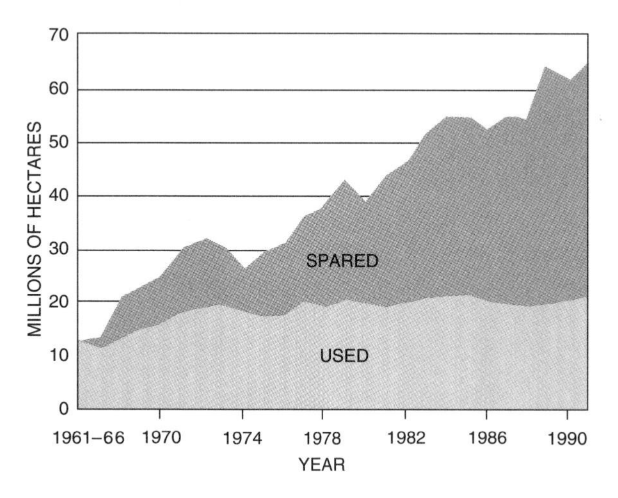


**Figure 18.2 LAND SPARED in India increases with improvement in the yield of wheat varieties. The upper line (*orange*) represents the growing area that would have had to be dedicated to farming, assuming yields remained at 1960s' levels. Actual usage over three decades is shown in green.**

agriculture in developing nations. But social investments there need also to be coupled with continued research: the challenges for science in devising appropriate and effective farm management practices in the developing world are still immense.

## New Farm Implements

Agriculture of the 21st century will see a range of novel diagnostic tools that carry the power of modern science to all levels of decision making. Biological test kits should become available for identifying viruses and other diseases on the spot, bringing sophisticated biotechnology right into farmers' fields. Although such innovations will first appear in high-income countries, we see expanding opportunities for their application in poorer nations as well, both in guiding research and, ultimately, in helping farmers manage their crops.

A noteworthy advance in agricultural technology will be expert computer systems that combine knowledge from many disciplines to help guide farmers' actions. Several such "decision support systems" are now under development to aid farmers with soil management. Most advanced are systems for handling phosphorus deficiency and high acidity, but others are in the offing. These computer applications not only present an expert analysis of what is known and the effect a certain management approach might produce, but they also can identify key gaps in knowledge that might be addressed by further data collection.

The strength of expert systems is that they can combine scientific principles with indigenous farming knowledge. A weakness is that they may not provide reliable answers for situations that evolve rapidly. Another approach, involving numerical simulations, can better deal with continuously changing effects such as weather. The aim now is to integrate expert systems and simulation models. As this goal is accomplished, farmers will increasingly use computer-based tools to help control pests, reduce water consumption and manage many different kinds of crops.

By fashioning new tools for farmers, science will have much to say about how sustainable agriculture will be practiced over the next several decades. In developed countries, scientific advances will bring greater productivity and lower costs, will help protect land and water and thus will ensure adequate stocks of food and fiber for the next century. In poorer countries, science-based technologies will undoubtedly be harder to introduce. Ultimately, however, they will be even more crucial in coping with marginal farm productivity and environmental destruction—and with poverty itself.

# Commentary: Outline for an Ecological Economy

*Countries can indeed prosper while protecting their environment.*

• • •

Heinrich von Lersner

Charged with looking after the health of Germany's land, air and water, I have discovered what anyone involved with the environment knows: it is difficult to predict the future. Nevertheless, it seems safe to forecast that advancements in technology will prove crucial for protecting the planet during the 21st century. And, most probably, Germany will remain among those countries leading the movement to find and apply such solutions.

Perhaps foremost in aiding the environment will be improvements in the generation and conversion of energy. Here the state of the art has been advancing rapidly. For instance, one now finds—even in central Europe—houses that can derive their energy exclusively from the sun, and it would seem likely that the trend toward self-sufficient dwellings will continue. Alongside that progress, improvements in transportation should drastically reduce the number of individually owned motor vehicles on the roads of developed nations. That change will further decrease overall energy consumption.

Like petroleum supplies today, freshwater will become a highly cherished commodity. Thus, new methods will be crafted so that industry can make do with a smaller amount of water by contaminating it less and recycling it more. Textile manufacturers in Germany have, for example, adopted equipment to filter chemical additives from wastewater. Because the filtrate and water are both reused, the effort not only helped the environment, it also proved economical enough to pay for the pollution-control equipment in 18 months.

But the real challenge in the next century will not be so much with industry as with agriculture. Better devices to desalinize water, transport it long distances and irrigate with it while avoiding wasteful evaporation will be particularly critical for people living in arid regions. Indeed, such technical advances in water management may help recover and protect arable land throughout the world.

Just as with the earth's land and water, the atmosphere needs to be scrupulously guarded. Efforts in that direction are in their formative stages. For example, the signatories of the Rio Convention accords limiting greenhouse gas emissions to the atmosphere met in the spring of 1995 in Berlin to move the treaty forward. Although the group did not forge satisfactory international law, those discussions gave increased impetus to finding means that might stabilize carbon dioxide output in the developed world and prevent its rapid growth in less industrialized countries.

Technical improvements should indeed help protect the environment from continued assault, but they must also serve to remedy past neglect. In Germany, as in many other developed countries, society has inherited substantial burdens: dangerous wastes were often handled improperly. Much of our land is afflicted with chemicals that it cannot break down, and these substances are now polluting our food and water. In the past few years the German government has tried to address this legacy by developing a number of techniques for cleaning contaminated soil, including biological and thermal methods to degrade toxins.

Although successful in many respects, efforts to clean up waste sites frequently prove to be slow and expensive. Our strategy for the next millennium will therefore be to avoid creating such difficulties wherever possible. Substances destined for disposal in the 21st century will have to meet strict criteria so that they do not pose a danger to future generations. This concern may require that waste first be treated—for example, by thermal processes that combine carbonization at low temperatures and incineration at high temperatures. Such advanced thermal methods are soon to be applied in Germany.

But better disposal schemes alone will not suffice to solve the fundamental problem of waste. Manufacturers will have to produce items that ensure, from start to finish, an environmentally compatible cycle. Advanced methods, such as "near net shape manufacturing," will cut waste by limiting the amount of machining needed for specialized parts. Complex items will be built to be separated neatly and easily into their components when the product is retired from use, and those pieces will be marked so that they can be identified electronically to facilitate their recycling. The technology for the separation of mixed materials will develop further, and factory production will become more and more efficient in its use of recycled materials. One can expect to see this evolution toward a closed-cycle economy accelerate as manufacturers are held legally and financially responsible for the ultimate disposability of their products—a revolutionary concept that is now beginning to be implemented in our country.

The situation in Germany serves to demonstrate that the most economically and ecologically successful nations today are those with liberal, market-oriented economies. But these places will continue to prosper only if industrialists, merchants and consumers all make it a priority to act in an ecologically responsible manner. Indeed, advanced countries that can maintain continued economic growth along with careful environmental stewardship in the 21st century will serve as compelling examples for the remainder of the world.

With the increasing human population and its demand for resources, environmental protection will face a host of challenges. But, to borrow a sentiment from Russell Train, former administrator of the U.S. Environmental Protection Agency, the greatest demand in the future will not be for coal, oil or natural gas; it will be for the time we need to adapt our laws, behaviors and technologies to the new requirements.

# Commentary: Outline for an Ecological Economy

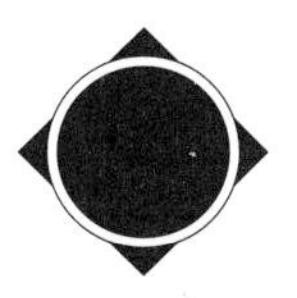

*Countries can indeed prosper while protecting their environment.*

• • •

Heinrich von Lersner

Charged with looking after the health of Germany's land, air and water, I have discovered what anyone involved with the environment knows: it is difficult to predict the future. Nevertheless, it seems safe to forecast that advancements in technology will prove crucial for protecting the planet during the 21st century. And, most probably, Germany will remain among those countries leading the movement to find and apply such solutions.

Perhaps foremost in aiding the environment will be improvements in the generation and conversion of energy. Here the state of the art has been advancing rapidly. For instance, one now finds—even in central Europe—houses that can derive their energy exclusively from the sun, and it would seem likely that the trend toward self-sufficient dwellings will continue. Alongside that progress, improvements in transportation should drastically reduce the number of individually owned motor vehicles on the roads of developed nations. That change will further decrease overall energy consumption.

Like petroleum supplies today, freshwater will become a highly cherished commodity. Thus, new methods will be crafted so that industry can make do with a smaller amount of water by contaminating it less and recycling it more. Textile manufacturers in Germany have, for example, adopted equipment to filter chemical additives from wastewater. Because the filtrate and water are both reused, the effort not only helped the environment, it also proved economical enough to pay for the pollution-control equipment in 18 months.

But the real challenge in the next century will not be so much with industry as with agriculture. Better devices to desalinize water, transport it long distances and irrigate with it while avoiding wasteful evaporation will be particularly critical for people living in arid regions. Indeed, such technical advances in water management may help recover and protect arable land throughout the world.

Just as with the earth's land and water, the atmosphere needs to be scrupulously guarded. Efforts in that direction are in their formative stages. For example, the signatories of the Rio Convention accords limiting greenhouse gas emissions to the atmosphere met in the spring of 1995 in Berlin to move the treaty forward. Although the group did not forge satisfactory international law, those discussions gave increased impetus to finding means that might stabilize carbon dioxide output in the developed world and prevent its rapid growth in less industrialized countries.

Technical improvements should indeed help protect the environment from continued assault, but they must also serve to remedy past neglect. In Germany, as in many other developed countries, society has inherited substantial burdens: dangerous wastes were often handled improperly. Much of our land is afflicted with chemicals that it cannot break down, and these substances are now polluting our food and water. In the past few years the German government has tried to address this legacy by developing a number of techniques for cleaning contaminated soil, including biological and thermal methods to degrade toxins.

Although successful in many respects, efforts to clean up waste sites frequently prove to be slow and expensive. Our strategy for the next millennium will therefore be to avoid creating such difficulties wherever possible. Substances destined for disposal in the 21st century will have to meet strict criteria so that they do not pose a danger to future generations. This concern may require that waste first be treated—for example, by thermal processes that combine carbonization at low temperatures and incineration at high temperatures. Such advanced thermal methods are soon to be applied in Germany.

But better disposal schemes alone will not suffice to solve the fundamental problem of waste. Manufacturers will have to produce items that ensure, from start to finish, an environmentally compatible cycle. Advanced methods, such as "near net shape manufacturing," will cut waste by limiting the amount of machining needed for specialized parts. Complex items will be built to be separated neatly and easily into their components when the product is retired from use, and those pieces will be marked so that they can be identified electronically to facilitate their recycling. The technology for the separation of mixed materials will develop further, and factory production will become more and more efficient in its use of recycled materials. One can expect to see this evolution toward a closed-cycle economy accelerate as manufacturers are held legally and financially responsible for the ultimate disposability of their products—a revolutionary concept that is now beginning to be implemented in our country.

The situation in Germany serves to demonstrate that the most economically and ecologically successful nations today are those with liberal, market-oriented economies. But these places will continue to prosper only if industrialists, merchants and consumers all make it a priority to act in an ecologically responsible manner. Indeed, advanced countries that can maintain continued economic growth along with careful environmental stewardship in the 21st century will serve as compelling examples for the remainder of the world.

With the increasing human population and its demand for resources, environmental protection will face a host of challenges. But, to borrow a sentiment from Russell Train, former administrator of the U.S. Environmental Protection Agency, the greatest demand in the future will not be for coal, oil or natural gas; it will be for the time we need to adapt our laws, behaviors and technologies to the new requirements.

# PART VI
# LIVING WITH NEW TECHNOLOGIES

# Technology Infrastructure

*Industrial advances will depend on setting new standards.*

• • •

Arati Prabhakar

The dictionary defines "infrastructure" as an underlying foundation. In society, that means such basic installations as roads, power grids and communications systems. It is the stuff we take for granted, at least when it works. Because this support is often out of sight, its essential role tends to be out of mind. For the most part, it should be. As technology advances, however, the infrastructure must evolve as well. Thus, now is the time to think about the enabling tools and underpinning technologies that will be needed in the next century.

The information revolution and global marketplace are demanding an increasingly diverse array of infrastructural technologies. When the National Bureau of Standards—now the National Institute of Standards and Technology (NIST)—was founded in the U.S. at the turn of the century, its mission was to create measurement standards so that items such as automobiles could be mass-produced most efficiently. For the automotive industry today, such guidelines pertain to far more than the sizes of interchangeable parts. To analyze properly the chemical composition of a car's exhaust, standardized samples of carbon monoxide and nitrous oxides, among other gases, are needed. So, too, thermocouples, which feed temperature readings into a car's microprocessor-based engine controller, must be made according to strict specifications so that their signals are accurate.

In a decade the automotive infrastructure will have expanded even more. Cars will very likely be designed and manufactured using standardized product data exchange specifications (PDES). Over computer networks, these digital blueprints will pass like relay batons between designers and engineers, making it easier to simulate the performance of sundry auto parts before they are made. Standardized formats for these specifications should also allow for more agile manufacturing practices, making it economically feasible to produce more custom-tailored models.

In addition, future cars may contain more parts made from composite materials, including mixes of polymers and ceramic reinforcements. These substitutes are as strong as steel yet much lighter. Hence, they could yield highly fuel-efficient, clean vehicles. Such materials were originally devised for defense applications; at the moment, they are too expensive to use in large amounts in automotive manufacturing. But the Advanced Technology Program (ATP)—unfolding at many high-tech companies in partnership with NIST—aims to develop affordable, high-performance varieties of composites. Companies that are arch rivals outside the ATP framework are now collaborating toward this end. ATP is enabling the industry as a whole to pursue this promising technology, which is too risky for any one company to take on.

Looking even further ahead, a variety of powerful sensors, computers and communications devices may innervate cars, roads, bridges and traffic management systems. Some of these additions may result from new chip designs that will themselves require new infrastructural technologies. Microprocessor chips, for example, will soon have parts only about 0.25 micron in size, putting them in the realm of large viruses. Innovations in microlithography and other microfabrication techniques should soon lower that scale to 0.1 micron or less—at which point new measuring devices will be needed.

NIST has already begun testing one such device, the Molecular Measuring Machine, or $M^3$. This instrument can map out subatomic detail over an area the size of a credit card. Equipped with the $M^3$, semiconductor manufacturers will be able to trace their measurements to references that are accurate to within less than 2.5 nanometers (or the width of about eight water molecules in a row). Such precision will assist in continuing efforts to shrink the size of integrated circuits and to increase the power of devices that contain them.

A team at NIST has worked on the hardware and software for the $M^3$ since 1987. To minimize errors caused by vibrations or temperature changes, the tiny probe at the heart of the instrument—a sophisticated scanning tunneling microscope—is housed within a basketball-size copper sphere; this sphere is then nested within a series of successively larger shells. A computer system uses laser interferometers, a meticulously machined sliding carriage and piezoelectric flexing elements to produce controllable displacements as small as 0.075 nanometer (or less than the diameter of a hydrogen atom). To validate the machine's performance for these minute motions, the team will turn to a nanoruler: an atomically smooth surface of a crystal such as tungsten diselenide. The accurately measured distance between single atoms in this crystal can serve as the ruler's gradations. Also, because the crystal is flat over an uncommonly large area, it can serve as an ultrahigh-accuracy geometry reference for $M^3$, much as a square does for a machinist.

On long road trips, your children might be entertained by electronic and communications devices manufactured with the help of $M^3$. To that end, the infrastructure of the coming century will probably include digital video standards so that interactive programs can be broadcast over complex information networks. Another focused program within the ATP is bringing together the many different players who hope to make this scenario real.

The list goes on. Workers are trying to devise fingerprint- or face-recognition systems that could allow you to enter your car quickly and easily without a key. Such systems will call for sophisticated software algorithms, however: they will need to recognize the same fingerprint, for example, even when its appearance varies slightly from one press to the next. Vendors and buyers will need standard benchmarks to compare the performance of these future software programs. Benchmarks will also be needed to judge the quality of laboratories doing genetic testing and other biotechnology-based analyses. Such procedures will probably be far more commonplace in 21st-century clinical settings than they are now.

Adam Smith wrote two centuries ago in *The Wealth of Nations* that the state is responsible for "erecting and maintaining those public institutions and those public works, which though they may be in the highest degree advantageous to a great society, are, however, of such a nature, that the profit could never repay the expense to any individual or small number of individuals, and which it, therefore, cannot be expected that any individual or small number of individuals should erect or maintain." These words have never been more true.

# Designing the Future

*Too frequently, product designers disregard the psychology of the user.*

• • •

Donald A. Norman

The difficulty of programming videocassette recorders has become a worldwide joke. "I'm a rocket scientist," one engineer complained to me. "I design missile systems, but I can't figure out how to program my VCR." Why is it that we sometimes have so much trouble working apparently simple things, such as doors and light switches, water faucets and thermostats, to say nothing of computers and automated factory equipment? The answer lies not with the hapless user but with designers who fail to think about products from the operator's point of view. The steps required to run modern devices frequently seem arbitrary and capricious often because they are indeed confusing.

Although most problems arise with electronic equipment, certain fundamental design flaws can be illustrated with simple mechanical objects. Consider the door. With most doors, there are only two possible actions: push or pull. But which? Where? Poorly designed doors turn the operation into a guessing game, requiring the posting of signs to indicate the appropriate action. Now suppose a door had a flat metal panel along one side. The panel by itself would essentially say, "Push here." You would immediately know how to proceed, because the maker included a visible cue to the door's operation. The best cues offer an intuitive indication of the things you can do with an object—what James J. Gibson of Cornell University had termed the object's "affordances." In general, if a simple piece of equipment such as a door or a kitchen stove requires labeling, that need is a sign of design failure. Wonderful capabilities become meaningless if they are hard to discover and use.

Providing unambiguous cues to the operation of a device is only one part of good design. A few other, related principles need to be invoked as well. First, people can manipulate things better when they understand the logic behind how the objects work. Designers can help convey this information by giving users a "conceptual model," or a simple way to think about how the device operates. For example, the modern computer often labels stored information as being in files and folders, as if our central processing units contained metal cabinets in which manila files were stuffed into hanging green folders. Of course, there are no physical files or folders inside the computer, but this model helps users understand how to save and retrieve their work.

Second, each operation should be followed promptly by some sort of feedback that indicates the operation was successful, even in cases where the output is not immediately apparent. The spinning clock or hourglass displayed by some computer systems is useful for indicating that a command was understood but that its instructions will take some time to complete.

Finally, the controls on a machine should be positioned in a way that correlates with their effects. On well-designed stoves, for example, if the burners are arranged in a rectangular pattern, the controls should also be arranged in a rectangular pattern, so the left rear control operates the left rear burner, and so on. Today most stoves have the burners arranged in a rectangle with their controls in a line: no wonder people frequently make mistakes, despite the labels.

As automation increases, the need to apply such principles becomes more urgent. Once upon a time, technology was mostly mechanical. Everything was built of levers, gears, cogs and wheels. Workers who operated tools could view many of the parts and could see the effects of their actions. People had some hope of understanding how large machinery and small gadgets worked, because the parts were visible. The operation of modern machines and the concepts behind their design are invisible and abstract. There may be nothing to see, nothing to guide understanding. Consequently, workers know less and less about the inner workings of the systems under their control, and they are at an immediate disadvantage when trouble erupts.

Such alienation has startling effects: most industrial and aviation accidents today are attributed to human error. When the majority of accidents stem from mistakes made by operators, the finding is a sign that the equipment is not designed appropriately for the people who must use it. Many manufacturers—and much of society—still follow the "blame and train" philosophy: when an accident occurs, blame the operators and retrain them. A more appropriate response would be to redesign devices in a way that minimizes the chance for error in the first place. And when errors do occur, the machinery should ensure that the mistakes are readily caught and corrected before they do damage. Most technologists do not have the proper training or knowledge needed to design such error-resistant systems. To cope with this gap, a discipline in applied cognitive science—variously called human factors, ergonomics or cognitive engineering—has arisen. Scientists in this field develop design concepts emphasizing the mental rather than physical side of design.

As the chapters in this book attest, we are in the midst of a sweeping technological transformation. But this revolution is also a human and social one. The great promised advances in knowledge, communications, cooperative work, education and entertainment will come about only if the technology truly fits the needs and capabilities of its users. To make technology that fits human beings, it is necessary to study human beings. But now we tend to study only the technology. As a result, people are required to conform to technology. It is time to reverse this trend, time to make technology conform to people.

# Digital Literacy

*Multimedia will require equal facility in word, image and sound.*

• • •

Richard A. Lanham

The word "literacy," meaning the ability to read and write, has gradually extended its grasp in the digital age until it has come to mean the ability to understand information, however presented. Increasingly, information is being offered in a new way: instead of black letters printed on a white page, the new format blends words with recorded sounds and images into a rich and volatile mixture. The ingredients of this combination, which has come to be called multimedia, are not new, but the recipe is.

New, too, is the mixture's intrinsic volatility. Print captures utterance—the words are frozen on the page. This fixity confers authority and sometimes even timeless immortality. That is why we value it, want to get things down in "black-and-white," write a sonnet, in Horace's words, "more lasting than bronze." The multimedia signal puts utterance back into time: the reader can change it, reformat and rescale it, transform the images, sounds and words. And yet, at the end of these elegant variations, the original can be summoned back with a keystroke.

Print literacy aimed to pin down information; multimedia literacy couples fixity and novelty in a fertile oscillation. Contrary to the proverbial wisdom, in a digital universe you can eat your cake and have it, too: keep your original and digest it on your own terms. And because digital code is replicable without material cost, you can give your cake away as well.

Printed books created the modern idea of "intellectual property" because they were fixed in form and difficult to replicate. One could therefore sell and own them, and the livelihoods of printer and author could be sustained. This copyright structure dissolves when we introduce the changeable multimedia signal. We will have to invent another scaffolding to fit the new literacy. Judging from the early signs, it won't be easy.

There is one other way in which digital flexibility is radical. If we ask, looking through the wide-angle lens of Western cultural history, "What does multimedia literacy do?", a surprisingly focused answer comes back. It recaptures the expressivity of oral cultures, which printed books, and handwritten manuscripts before them, excluded.

In writing this text, for example, I have been trying to create a credible "speaking voice," to convince you that I am a person of sense and restraint. Now imagine that you can "click" on an "author box." I appear as a moving image, walk into the margin and start to speak, commenting on my own argument, elaborating it, underlining it with my voice, gesture and dress—as can happen nowadays in a multimedia text.

What has changed? Many of the clues we use in the oral culture of daily life, the intuitive stylistic judgments that we depend on, have returned. You can see me for yourself. You can hear my voice. You can feed that voice back into the voiceless prose and thus animate it. Yet the writing remains as well. You can see the author with stereoscopic depth, speaking in a space both literate and oral.

Oral cultures and literate cultures go by very different sets of rules. They observe different senses of time, as you will speedily understand if you listen to one of Fidel Castro's four-hour speeches. Oral cultures prolong discourse because, without it, they cease to be; they exist only in time. But writing compresses time. An author crams years of work into some 300 pages that the reader may experience in a single day.

Oral and literate cultures create different senses of self and society, too. The private reflective self created by reading differs profoundly from the unselfconscious social role played by participants in a culture that knows no writing. Literacy allows us to see human society in formal terms that are denied to an oral culture that just plays out its drama.

The oral and written ways of being in the world have contended rancorously throughout Western history, the rancor being driven more often than not by literate prejudice against the oral rules. Now the great gulf in communication and in cultural organization that was opened up by unchanging letters on a static surface promises to be healed by a new kind of literacy, one that orchestrates these differences in a signal at the same time more energizing and more irenic than the literacy of print.

If we exchange our wide-angle cultural lens for a close-up, we can observe the fundamental difference between the two kinds of literacies. In the world of print, the idea and its expression are virtually one. The meaning takes the form of words; words generate the meaning. Digital literacy works in an inherently different way. The same digital code that expresses words and numbers can, if the parameters of expression are adjusted, generate sounds and images. This parametric variation stands at the center of digital expressivity, a role it could never play in print.

The multiple facets of this digital signal constitute the core difference between the two media, which our efforts in data visualization and sonification have scarcely begun to explore. If we think of the institutional practices built on the separation of words, images and sounds—such as separate departments for literature, art and music—we can glimpse the profound changes that will come when we put them back together.

To be deeply literate in the digital world means being skilled at deciphering complex images and sounds as well as the syntactical subtleties of words. Above all, it means being at home in a shifting mixture of words, images and sounds. Multimedia literacy makes us all skilled operagoers: it requires that we be very quick on our feet in moving from one kind of medium to another. We must know what kinds of expression fit what kinds of knowledge and become skilled at presenting our information in the medium that our audience will find easiest to understand.

We all know people who learn well from books and others who learn by hands-on experience; others, as we say in music, "learn by ear." Digital literacy greatly enhances our ability to suit the medium both to the information being offered and to the audience. Looked at one way, this new sensory targeting makes communication more efficient. Looked at another, it simply makes it more fun.

The multimedia mixture of talents was last advanced as an aristocratic ideal by the Renaissance humanists. The courtly lord and lady were equally accomplished in poetry, music and art. The Renaissance ideal now presents itself, broadened in scope and coarsened in fiber perhaps, as the common core of citizenship in an information society.

At its heart, the new digital literacy is thus profoundly democratic. It insists that the rich mixture of perceptive talents once thought to distinguish a ruling aristocracy must now be extended to everyone. It thus embodies fully the inevitable failures, and the extravagant hope, of democracy itself.

# The Information Economy

*How much will two bits be worth*
*in the digital marketplace?*

• • •

Hal R. Varian

Advances in computers and data networks inspire visions of a future "information economy" in which everyone will have access to gigabytes of all kinds of information anywhere and anytime. But information has always been a notoriously difficult commodity to deal with, and, in some ways, computers and high-speed networks make the problems of buying, selling and distributing information goods worse rather than better.

To start with, the very abundance of digital data exacerbates the most fundamental constraint on information commerce—the limits of human comprehension. As Nobel laureate economist Herbert A. Simon puts it: "What information consumes is rather obvious: it consumes the attention of its recipients. Hence a wealth of information creates a poverty of attention, and a need to allocate that attention efficiently among the overabundance of information sources that might consume it." Technology for producing and distributing information is useless without some way to locate, filter, organize and summarize it. A new profession of "information managers" will have to combine the skills of computer scientists, librarians, publishers and database experts to help us discover and manage information. These human agents will work with software agents that specialize in manipulating information—offspring of indexing programs such as Archie, Veronica and various "World Wide Web crawlers" that aid Internet navigators today.

The evolution of the Internet itself poses serious problems. Now that the Internet has been privatized, several companies are competing to provide the backbones that will carry traffic between different local networks, but workable business models for interconnection—who pays how much for each packet transmitted, for example—have yet to be developed. If interconnection standards are developed that make it cheap and easy to transmit information across independent networks, competition will flourish. If technical or economic factors make interconnection difficult, so that transmitting data across multiple networks is expensive or too slow, the largest suppliers can offer a significant performance advantage; they may be able to use this edge to drive out competitors and monopolize the market.

Similar problems arise at the level of the information goods themselves. There is a growing need for open standards for formats used to represent text, images, video and other collections of data, so that one producer's data will be accessible to another's software. As with physical links, it is not yet clear how to make sure companies have the right economic incentives to negotiate widely usable standards.

In addition to standards for the distribution and manipulation of information, we must develop standards for networked economic transactions: the actual exchange of money for digital goods.

There are already more than a dozen proposals for ways to conduct secure financial transactions on the Internet. Some of them, such as the DigiCash system, involve complex encryption techniques; others, such as that used by First Virtual, are much simpler. Many of these protocols are implemented entirely in software; others enlist specialized hardware to support electronic transactions. "Smart" credit cards with chips embedded in them can perform a variety of authentication and accounting tasks.

Even when the financial infrastructure becomes widely available, there is still the question of how digital commodities will be priced. Will data be rented or sold? Will articles be bundled together, as is done today in magazines and newspapers, or will consumers purchase information on an article-by-article basis? Will users subscribe to information services, or will they be able to buy data spontaneously? How will payment be divided among the various parties involved in the transaction, such as authors, publishers, libraries, on-line services and so on? Not one of these questions has a definitive answer, and it is likely that many market experiments will fail before viable solutions emerge.

The shared nature of information technology makes it critical to address issues of standardization and interoperability sooner rather than later. Each consumer's willingness to use a particular piece of technology—such as the Internet—depends strongly on the number of other users. New communications tools, such as fax machines, VCRs and the Internet itself, have typically started out with long periods of relatively low use followed by exponential growth, which implies that changes are much cheaper and easier to make in the early stages. Furthermore, once a particular technology has penetrated a significant portion of the market, it may be very difficult to dislodge. Fortunes in the computer industry have been made and lost from the recognition that people do not want to switch to a new piece of hardware or software—even if it is demonstrably superior—because they will lose both the time they have invested in the old ways and the ability to share data easily with others. If buyers, sellers and distributors of information goods make the wrong choices now, repairing the damage later could be very costly.

This discussion about managing, distributing and trading in information is overshadowed by the more fundamental issue of how much data authors and publishers will be willing to make available in electronic form. If intellectual property protection is too lax, there may be inadequate incentives to produce new electronic works; conversely, if protection is too strict, it may impede the free flow and fair use of information. A compromise position must be found somewhere between those who suggest that all information should be free and those who advocate laws against the electronic equivalent of browsing at a magazine rack.

Extending existing copyright and patent law to apply to digital technologies can only be a stopgap measure. Law appropriate for the paper-based technology of the 18th century will not be adequate to cope with the digital technology of the 21st; already the proliferation of litigation over software patents and even over the shape of computer-screen trash cans makes the need for wholesale revisions apparent.

Computer scientists have been investigating various forms of copy protection that could be used to enforce whatever legal rules may be put into place. Although such protection often inconveniences users and requires additional hardware and software, ubiquitous network access and more powerful machines may eventually allow for unobtrusive and effective protection. File servers, for example, can track who owes how much to whom for the use of particular information, and documents can be subtly encoded so that investigators can trace the provenance of illicit copies.

Faced with such a daunting list of problems, one might be led to question whether a viable information economy will ever take shape, but I believe there are grounds for optimism. During the 1980s, 28,000 for-profit information libraries sprang up in the U.S. alone. Every week more than 50 million people visit these facilities, where they can rent 100 gigabytes of information for only two or three dollars a day.

Although these video rental stores faced many of the same problems of standards, intellectual-property protection, and pricing that the Internet faces today, the industry grew from nothing to $10 billion a year in only a decade. Ten years from now we may find the economic institutions of the information economy a similarly unremarkable part of our day-to-day life.

# The Emperor's New Workplace

*Information technology evolves more quickly than behavior.*

• • •

Shoshana Zuboff

According to the U.S. Department of Commerce, 1990 was the first year capital spending on the information economy—that is, on computers and telecommunications equipment—exceeded capital spending on all other parts of the nation's industrial infrastructure. Scholars and commentators have cited these data as evidence the U.S. economy is now firmly rooted in the information age. They routinely declare that an "information economy" has replaced the industrial economy that dominated most of the 20th century. I heartily dissent.

In a true information economy, information is the core resource for creating wealth. Constructing such an economy demands more than just a proliferation of computers and data networks. It requires a new moral vision of what it means to be a member of an organization and a revised social contract that binds members of a firm together in ways radically different from those of the past. So far patterns of morality, sociality and feeling are evolving much more slowly than technology. Yet without them, the notion of an information economy is much like the foolish emperor of the fairy tale, naked and at risk.

A historical perspective makes the problem clearer. Early in the 20th century an organizational form—the functional hierarchy—was invented to meet the business challenges of increasing throughput and lowering unit costs. Business processes were divided into separate functions—manufacturing, engineering, sales and so on. Other innovative features included mass-production techniques, the minute fragmentation of tasks, the professionalization of management, the growth of the managerial hierarchy to standardize and control operations, and the simplification and delegation of administrative functions to a newly contrived clerical workforce. Collectively, these components were incredibly successful; they came to define the modern workplace.

The industrial hierarchy rested on the premise that complexity could constantly be removed from lower level jobs and passed up to the management ranks. That is, clerks and factory workers became progressively less involved in the overall business of a firm as their jobs were narrowed and stripped of opportunities to exercise judgment. Automation was a primary means of accomplishing this. Meanwhile the manager's role evolved as guardian of the organization's centralized knowledge base. His legitimate authority derived from being credited as someone fit to receive, interpret and communicate orders based on the command of information.

We have come to accept that a managerial hierarchy operating in this way reflects a reasonable division of labor. We are less comfortable discussing the moral vision at its heart, something I call the "division of love." I suggest that the managerial hierarchy drew life not only from considerations of its efficiency but also from the ways in which some members of the organization were valued and others devalued.

In the brave new age of the information economy, this system cannot hold. Mass-market approaches have been forced to give way to a highly differentiated and often information-saturated marketplace in which firms must distinguish themselves through the value they add in response to customers' priorities. Information technologies now provide the means for generating such value with speed and efficiency.

Doing so means using the modern information infrastructure to cope with the complexities of a business outside a central managerial cadre. It is more efficient to handle complexity wherever and whenever it first enters the organization—whether during a sale, during delivery or in production.

This approach is now possible because of the way the unique characteristics of information technologies can transform work at every organizational level. Initially, most people regarded computers in the workplace as the next phase of automation. But whereas automation effectively hid many operations of the overall enterprise from individual workers, information technology tends to illuminate them. It can quickly give any employee a comprehensive view of the entire business or nearly infinite detail on any of its aspects.

I coined the word "informate" to describe this action. These technologies informate as well as automate: they surrender knowledge to anyone with the skills to access and understand it. Earlier generations of machines decreased the complexity of tasks. In contrast, information technologies can increase the intellectual content of work at all levels. Work comes to depend on an ability to understand, respond to, manage and create value from information. Thus, efficient operations in the informated workplace require a more equitable distribution of knowledge and authority. The transformation of information into wealth means that more members of the firm must be given opportunities to know more and to do more.

To avail themselves of the opportunities, firms must be prepared to drive a stake into the heart of the old division of labor (and the division of love sustaining it). Exploiting the informated environment means opening the information base of the organization to members at every level, assuring that each has the knowledge, skills and authority to engage with the information productively. This revamped social contract would redefine who people are at work, what they can know and what they can do.

The successful reinvention of the firm consistent with the demands of an information economy will continue to be tragically limited as long as the principal features of modern work are preserved. Unlocking the promise of an information economy now depends on dismantling the very same managerial hierarchy that once brought greatness. Only then can the emperor come in from the cold, because we will have found the way to clothe him.

# What Technology Alone Cannot Do

*Technology will not provide us all with health, wealth and big TVs.*

• • •

Robert W. Lucky

The subway sways and creaks as it travels away from Manhattan on the elevated tracks through Queens. Looking at the passing cityscape, I see the familiar skeletal steel of the Unisphere rising above the apartment buildings like some dark moon. As the subway rushes toward that rusting remnant of the 1964 World's Fair, I am transported in memory back to my eager visits there as a young technologist. I see again through youthful eyes the excitement and the promises for the future made in the exhibitions of that now demolished fair.

Surely every reader remembers some similar experience—an exposition, an exhibit or a theme park portraying a glittering technological future, where smiling people clustered around large television sets in solar homes that required no maintenance. As this standard demographic family basked in the glow of mindless electronic entertainment, "smart" toasters and robot vacuum cleaners hummed subserviently in the background. Even the standard demographic dog watched attentively, sporting a slight smile of superiority.

I remember the vision of a future in which drudgery had been eliminated, where everyone had health and wealth and where our chief preoccupation had become filling the void of expanding leisure time. Life had become effortless and joyful, and science and technology had made it all possible. Like most visitors, I suppose, I was caught up in the euphoria of that vision and believed in it completely.

The subway rounds a bend, and the sudden jolt brings me back to the present. I am once again enveloped in the microcosm of contemporary society randomly gathered in the drab confines of the rushing car. Where is that World's Fair family today, I wonder? What happened to those plastic people with their plastic home and their plastic lives? Surely none of my fellow passengers fit their description. These people look as though they have experienced continuous drudgery. The group seems to be divided into two nonoverlapping categories—those with no leisure time and those with nothing but leisure time. If technology was going to solve their problems, then technology has apparently failed.

Even as I consider that possibility, though, I vehemently reject it. What we accomplished in the three decades since that fair closed went far, far beyond the most outrageous projections we could have then conceived: we walked on the moon; we brought back pictures from the farthest reaches of the solar system and from orbiting telescopes peering into the very origins of the universe; we blanketed the earth with fiber-optic links and networked the planet with high-speed digital communications; we created microchips containing millions of transistors and costing so little that many homes could have a computer more powerful than the mainframes of that earlier day; we unraveled DNA and probed the fundamental building blocks of nature. No, science and technology did not fail. They just weren't enough.

There is a simplistic notion, which is crystallized in exhibitions such as World's Fairs, that we can invent the future. Alas, it does not seem to be so. Those awesome scientific developments of the past three decades have apparently missed my subway companions. Life's everyday problems, as well as the deeper problems of the human condition, seem resistant to quick technological fixes. The solutions shown in that forgotten World's Fair now appear at best naive or superficial, if not misleading or just plain wrong. Nevertheless, if you visit a similar exhibit today, I am sure it will acquire over time these same attributes.

If we could go back to 1964 and create in retrospect an exhibition of the future, what would we now include? Certainly the scientific accomplishments would deserve mention, but they would be framed in a social context. In our imaginary fair we would shock our disbelieving visitors by predicting the end of the cold war and the disintegration of the Soviet Union. We would say that nuclear missiles would cease to occupy people's fears but that, unfortunately, smaller wars and racial and ethnic strife would proliferate.

Sadly, we would have to predict that the inner cities would decay. We would report that a new disease of the immune system would sweep over the earth. We would tell of pollution cluttering the great cities and mention that environmental concerns would drive government policies and forestall the growth of nuclear power. Malnutrition, illiteracy and the gap between the haves and the have-nots would be as great as ever. Illegal drugs, terrorism and religious fundamentalism would become forces of worldwide concern. Only a small fraction of families would have two parents and a single income. Oh, and by the way, we would have big television sets.

Some years ago I was invited to be a guest on a television show hosted by a well-known, aggressive and sometimes offensive character. The host's producers assured me that this would be a serious show, marking the beginning of a new image for their client. The program would be devoted to a look at the future through the eyes of experts. Somehow, in spite of the firmly voiced apprehensions of my company's public relations people, I ended up in front of a television camera alongside scholars of education, medicine, finance, crime and the environment. I was "the technologist."

Somewhere I have a tape recording of that televised show, but I intend never to watch it. The educator told how illiteracy was on the rise and test scores were plummeting. The medical researcher said that progress in conquering the dread diseases was at a standstill. The financial expert forecasted that world markets would crumble. The criminologist gave statistics on the rise of crime, and the environmentalist predicted ubiquitous and unstoppable pollution. All agreed that the future would be bleak.

When the host finally turned to me, I said something to the effect that technology was neat, and it would make work easier and leisure time more fun. I think I said we all would have big television sets. I remember the way the other guests stared at me. "Can you believe such naiveté?" they seemed to say to one another. The host looked pained; he was into predictions of doom. Stubbornly, if feebly, I insisted that life would be better in the future because of technology.

Even today I blush remembering my ineptitude. But I do still believe there is a germ of truth in optimistic predictions. The continuous unraveling of nature's mysteries and the expansion of technology raise the level on which life, with all its ups and downs, floats. Science and technology, however, depend for their effect on the complex, chaotic and resistant fabric of society. Although they cannot in themselves make life better for everyone, they create a force that I believe has an intrinsic arrow, like time or entropy, pointing relentlessly in one direction: toward enhancing the quality of life.

I sometimes reflect on the historical contributions technology has made to human comfort. When I visit the ancient castles of Europe, I imagine the reverberant call of trumpets and the pageantry and glory that once graced those crumbling ruins.

But then I shiver in the dampness and cold and notice the absence of sanitation. Life now is unquestionably better, and there is no reason to think it will not be similarly improved in the future.

Overall progress is assured, but science and technology interwork with societal factors that determine their instantaneous utility and ultimate effect. This interplay is especially apparent in the current evolution of cyberspace. Ironically, the term was coined by William Gibson in *Neuromancer,* a novel that depicted a future in which the forces of computerized evil inhabited a shadowy world of networked virtual reality. Gibson's vision of gloom seems in step with those of my televised companions. In reality, though, cyberspace is a place where new communities and businesses are growing, and it seems largely to benefit its participants.

There are a multitude of meetings and conferences at which scientists and engineers talk about the evolution of the information infrastructure. But what do we talk about? Not the technology, to be sure. We talk about ethics, law, policy and sociology. Recognizing this trend, a friend recently wondered aloud if, since technologists now regularly debate legal issues, lawyers have taken to debating technology. At my next meeting with lawyers, I asked if this were indeed the case. They looked at me blankly. "Of course not," someone finally said. In fact, lawyers are just as comfortable in cyberspace as are scientists. It is a social invention. The problems that we all debate pertain to universal access, rights to intellectual property, privacy, governmental jurisdiction and so forth. Technology was the enabler, but these other issues will determine the ultimate worth of our work.

The Unisphere is receding from view, and my memories fade. As I look around the subway, I sense that my companions do not care about cyberspace or anything else so intangible. The never-ending straight track ahead and the relentless forward thrust of the car seem indicative of technology and life. Despite continuous motion on the outside, life on the inside seems still and unaffected. The Unisphere and the technology that it represents drift silently by, perceived only dimly through the clouded windows. The only real world—the one inside the car—remains unmoved in the midst of motion. Science urges us ever forward, but science alone is not enough to get us there.

# THE AUTHORS

**JOHN RENNIE** is editor in chief of SCIENTIFIC AMERICAN.

**DAVID A. PATTERSON** ("Microprocessors in 2020") has taught since 1977 at the University of California, Berkeley, where he now holds the E. H. and M. E. Pardee Chair in Computer Science. He is a member of the National Academy of Engineering and is a fellow of both the Institute of Electrical and Electronic Engineers and the Association for Computing Machinery. He has won several teaching awards, co-authored five books and consulted for many companies, including Digital, Intel and Sun Microsystems. His current research is on large-scale computing using networks of workstations.

**GEORGE I. ZYSMAN** ("Wireless Networks") is chief technical officer of AT&T's Network Wireless Systems business unit and a director of AT&T Bell Laboratories.

**DOUGLAS B. LENAT** ("Artificial Intelligence") has been working on the codification of commonsense knowledge for more than 10 years. He received his Ph.D. in computer science in 1976 from Stanford University with a demonstration that computers could be programmed to propose original mathematical theorems. He is now president of Cycorp, Inc., in Austin, Tex., and a consulting professor of computer science at Stanford.

**BRENDA LAUREL** ("Commentary: Virtual Reality"), an expert on interactive media, video games and virtual reality, is on the research staff of Interval Research Corporation in Palo Alto, Calif. She is the author of *Computers as Theatre* (Addison-Wesley, 1993).

**RUSSELL DAGGATT** ("Commentary: Satellites for a Developing World") is president of Teledesic Corporation.

**TONY R. EASTHAM** ("High-Speed Rail: Another Golden Age?") is a professor in the Department of Electrical and Computer Engineering at Queen's University in Kingston, Ontario. In 1995 he was visiting professor of transportation systems engineering at the University of Tokyo. His research interests include advanced ground transportation, electric drives and the design of machines and electromagnetic devices.

**DIETER ZETSCHE** ("The Automobile: Clean and Customized") is a member of the managing board of Mercedes-Benz AG, where he is responsible for sales, marketing and service. Previously, he oversaw passenger car development. Since joining the company in 1976, he has also held positions overseeing development of commercial vehicles in Brazil. He was appointed to the company's managing board in 1988 and became president of Freightliner Corporation in Portland, Ore., in 1991.

**EUGENE E. COVERT** ("Evolution of the Commercial Airliner") is the T. Wilson Professor of Aeronautics at the Massachusetts Institute of Technology and director of the university's Center for Aerodynamic Studies. After receiving a bachelor's and a master's degree in aeronautic engineering from the University of Minnesota, he went on to complete an Sc.D. in the same discipline. He served as chief scientist for the U.S. Air Force from 1972 to 1973 and headed the department of aeronautics and astronautics at M.I.T. from 1985 to 1990. He is a director of the American Institute of Aeronautics and Astronautics and sits on the board of several corporations, including AlliedSignal and Rohr, Inc.

**FREEMAN J. DYSON** ("21st-Century Spacecraft") has explored ideas in physics, engineering, politics, arms control, history and literature over his long career. His interest in space travel dates at least to age 9, when he wrote a short story about a manned mission to the moon; his classic book *Infinite in All Directions* discusses human destiny in space. Dyson worked as a boffin for Britain's Royal Air Force Bomber Command during World War II. Afterward, he completed his bachelor's degree at the University of Cambridge and moved to the U.S. Since 1953 he has held the position of professor of physics at the Institute for Advanced Study in Princeton, N.J.

**ROBERT CERVERO** ("Commentary: Why Go Anywhere?") is professor of city and regional planning at the University of California, Berkeley.

**W. FRENCH ANDERSON** ("Gene Therapy") is director of the Gene Therapy Laboratories and professor of biochemistry and pediatrics at the University of Southern California School of Medicine. He received an A.B. from Harvard College in 1958, an M.A. from Cambridge University in 1960 and returned to Harvard to complete his M.D. in 1963. Before moving to U.S.C., he spent 27 years at the National Institutes of Health, where he studied gene transfer and gene expression in mammals.

**ROBERT LANGER** and **JOSEPH P. VACANTI** ("Artificial Organs") have been collaborating for almost 20 years on the problems of organ failure and reconstructive surgery. Most recently, their attention has turned to tissue engineering. Langer focuses on the engineering aspects of the research; Vacanti is more concerned with the tissue. Langer, whose other major research interest is drug-delivery systems, is the Kenneth J. Germeshausen Professor of Chemical and Biomedical Engineering at the Massachusetts Institute of Technology. Vacanti is associate professor of surgery at Harvard Medical School and director of the Laboratory for Transplantation and Tissue Engineering at Children's Hospital.

**NANCY J. ALEXANDER** ("Future Contraceptives"), who holds a Ph.D. in physiology from the University of Wisconsin, is chief of the Contraceptive Development Branch in the Center for Population Research at the National Institute of Child Health and Human Development in Bethesda, Md., where she directs a wide variety of research projects aimed at developing fertility-regulating methods for men and women. She is also adjunct professor of obstetrics and gynecology at Georgetown University Medical Center and is on the steering committee of the World Health Organization's Task Force on Methods for the Regulation of Male Fertility.

**ARTHUR CAPLAN** ("Commentary: An Improved Future?") is director of the Center for Bioethics at the University of Pennsylvania Medical Center and founding president of the American Association of Bioethics. He received his Ph.D. in philosophy in 1979 from Columbia University, where he also studied genetics and evolutionary biology. He was a member of the President Bill Clinton's Health Care Task Force and serves on the President's Commission on the Gulf War Syndrome.

**GEORGE M. WHITESIDES** ("Self-Assembling Materials") is Mallinckrodt Professor of Chemistry at Harvard University. He received his Ph.D. in 1964 from the California Institute of Technology and taught at the Massachusetts Institute of Technology for almost 20 years before joining the chemistry department at Harvard in 1982. His interest in both materials science and biology and, in particular, his fascination with smaller and smaller structures drew him toward investigations of self-assembly.

**KAIGHAM J. GABRIEL** ("Engineering Microscopic Machines") manages a research program in microelectromechanical systems in the Electronic Systems Technology Office at the Department of Defense's Advanced Research Projects Agency. He received a doctorate in electrical engineering and computer science from the Massachusetts Institute of Technology in 1983. While a member of the robotic systems research department at AT&T Bell Laboratories from 1985 to 1991, he formed a research group that explored the use of microelectromechanical systems for data storage, manufacturing and optoelectronics.

**CRAIG A. ROGERS** ("Intelligent Materials") directs the Center for Intelligent Materials Systems and Structures at the Virginia Polytechnic Institute and State University. He is the founding editor in chief of the *Journal of Intelligent Materials Systems and Structures* and has achieved much recognition for his contributions to the field, including awards from the National Science Foundation and the Society for Mechanical Engineers.

**PAUL C. W. CHU** ("High-Temperature Superconductors"), who directs the Texas Center for Superconductivity at the University of Houston, earned his doctorate from the University of California, San Diego. He has served as a consultant to several organizations and received numerous awards, including the National Medal of Science. Besides studying superconductivity, he also researches magnetism and dielectric materials.

**JOSEPH F. ENGELBERGER** ("Commentary: Robotics in the 21st Century"), often called the father of robotics, is chairman of Transitions Research Corporation in Danbury, Conn., which seeks to develop personal-service robots.

**WILLIAM HOAGLAND** ("Solar Energy") received an M.S. degree in chemical engineering from the Massachusetts Institute of Technology. After working for Syntex, Inc., and the Procter & Gamble Company, he joined the National Renewable Energy Laboratory (formerly the Solar Energy Research Institute) in Golden, Colo., where he managed programs in solar materials, alcohol fuels, biofuels and hydrogen. He is president of W. Hoagland & Associates, Inc., in Boulder. The editors would like to acknowledge the assistance of Allan Hoffman of the Department of Energy in the preparation of this chapter.

**HAROLD P. FURTH** ("Fusion"), professor of astrophysical sciences at Princeton University, directed the Princeton Plasma Physics Laboratory from 1980 to 1990. After earning a Ph.D. in high-energy physics from Harvard University in 1960, he moved to Lawrence Livermore Laboratory and then to Princeton to work on magnetic fusion. Furth is a member of the National Academy of Sciences and the American Academy of Arts and Sciences; he has received, among other honors, the E. O. Lawrence Memorial Award from the Atomic Energy Commission. Furth acknowledges the assistance of Kevin M. McGuire, also at the Princeton Plasma Physics Laboratory, in preparing this chapter.

**ROBERT A. FROSCH** ("The Industrial Ecology of the 21st Century") is a senior research fellow at the John F. Kennedy School of Government at Harvard University. He has served as assistant executive director of the United Nations Environment Program and as administrator of the National Aeronautics and Space Administration. In 1993 he retired as vice president of General Motors Corporation, where he was in charge of the North American Operations Research and Development Center.

**DONALD L. PLUCKNETT** and **DONALD L. WINKELMANN** ("Technology for Sustainable Agriculture") have devoted their careers to advancing agriculture around the world. Between 1960 and 1980 Plucknett was professor of agronomy and soil science at the University of Hawaii, where he studied problems of tropical agriculture. He has worked for the U.S. Agency for International Development and for the World Bank and is now president of Agricultural Research and Development, a consulting firm. Winkelmann was professor of economics at Iowa State University before joining the International Maize and Wheat Improvement Center in 1971. He was director general of the center from 1985 until 1994 and now serves as chairman of the Technical Advisory Committee of the Consultative Group on International Agricultural Research.

**HEINRICH VON LERSNER** ("Commentary: Outline for an Ecological Economy") is president of Umweltbundesamt in Berlin, the German government equivalent of the U.S. Environmental Protection Agency.

**ARATI PRABHAKAR** ("Technology Infrastructure") is director of the National Institute of Standards and Technology in Gaithersburg, Md. She received her Ph.D. in applied physics from the California Institute of Technology in 1984.

**DONALD A. NORMAN** ("Designing the Future") is professor emeritus of cognitive science at the University of California, San Diego, and vice president of advanced technology at Apple Computer, Inc. Currently he is applying human-centered design principles to the design of computer systems.

**RICHARD A. LANHAM** ("Digital Literacy") is the author of several books on literary criticism and prose stylistics. His latest, *The Electronic Word*, was published simultaneously on paper and laser disc. He is professor emeritus of English at the University of California, Los Angeles, and president of Rhetorica, Inc., a media production company.

**HAL R. VARIAN** ("The Information Economy") is dean of the School of Information Management and Systems at the University of California, Berkeley, where he studies the emerging institutions of the electronic marketplace.

**SHOSHANA ZUBOFF** ("The Emperor's New Workplace") has written extensively about how computers will affect the future of work. She is Benjamin and Lillian Hertzberg Professor of Business Administration at Harvard Business School.

**ROBERT W. LUCKY** ("What Technology Alone Cannot Do") began working for Bell Laboratories in 1961 and has been vice president of applied research at Bellcore since 1992. He received his Ph.D. in electrical engineering in 1961 from Purdue University.

# BIBLIOGRAPHY


1. Microprocessors in 2020

Baskett, F., and J. L. Hennessy. 1993. Microprocessors: From desktop to supercomputers. *Science* 261 (August 13): 864–871.

Hennessy, J. L., and D. A. Patterson. 1994. *Computer organization and design: The hardware/software inter-face.* Morgan Kaufmann Publishers.

Patterson, D. A., and J. L. Hennessy. 1995. *Computer architecture: A quantitative approach,* 2d ed. Morgan Kaufmann Publishers.

Wilkes, M. V. 1995. *Computing perspectives.* Morgan Kaufmann Publishers.

2. Wireless Networks

Wireless technology. 1993. Special issue of *AT&T Technical Journal* 72 (July–August).

Wireless personal communications. 1995. Special issue of *IEEE Communications* 33 (January).

3. Artificial Intelligence

Lenat, Douglas B., and R. V. Guha. 1989. *Building large knowledge-based systems: Representation and inference in the CYC project.* Addison-Wesley.

Nilsson, Nils. 1995. Eye on the prize. *AI Magazine* 16 (Summer): 9–17.

Lenat, Douglas B. 1995. Steps to sharing knowledge. In *Toward very large knowledge bases,* ed., N. J. I. Mars. IOS Press.

4. High-Speed Rail: Another Golden Age?

Vranich, Joseph. 1993. *Supertrains: Solutions to America's transportation gridlock.* St. Martin's Press.

High Speed Rail/Maglev Association, Alexandria, Va. 1994. *The 21st century limited: Celebrating a decade of progress.* Reichman Frankle, Englewood Cliffs, N.J.

Harrison, J. A. 1995. High-speed ground transportation is coming to America—slowly. *Journal of Transportation Engineering* 121 (March–April): 117–123.

Eastham, Tony R. 1995. High–speed ground transportation development outside the United States. *Journal of Transportation Engineering* 121 (September–October): 411–416.

5. The Automobile: Clean and Customized

Seiffert, Ulrich, and Peter Walzer. 1991. *Automobile technology of the future.* Society of Automotive Engineers, Warrendale, Pa.

Riley, Robert Q. 1994. *Alternative cars in the 21st century: A new personal transportation paradigm.* Society of Automotive Engineers, Warrendale, Pa.

DeCicco, John, and Marc Ross. 1994. Improving automotive efficiency. *Scientific American* 271 (December): 52–57.

Deker, Uli. 1995. Riders on the storm. In *Daimler-Benz HighTech Report, March 1995.* Available from Daimler-Benz-Leserservice, Postfach 1271, D-73762 Neuhausen, Germany.

6. Evolution of the Commercial Airliner

Cutler, John. 1981. *Understanding aircraft structures.* Granada Publishing.

Shevel, Richard J. 1989. *Fundamentals of flight.* Prentice-Hall.

Aeronautics and Space Engineering Board and the National Research Council. 1992. *Aeronautical technologies for the twenty-first century.* National Academy Press.

McCormick, Barnes W. 1995. *Aerodynamics, aeronautics and flight mechanics.* John Wiley & Sons.

7. 21st-Century Spacecraft

Dyson, Freeman J. 1983. Science and space. In *The first 25 years in space,* ed., Allan Needell. Smithsonian Institution Press..

———. 1990. Major Observatories in Space. In *Observatories in Earth and beyond,* ed., Yoji Kondo. Kluwer Academic Publishers.

Staehle, Robert L., et al. 1993. *Pluto mission progress report: Lower mass and flight time through advanced technology insertion.* Report IAF-93-Q.5.410, 44th Congress of the International Astronautical Federation.

Kakuda, Roy, Joel Sercel and Wayne Lee. 1994. *Small body rendezvous mission using solar electric propulsion: Low-cost mission approach and technology requirements.* Report IAA-L-0710, Institute of Aeronautics and Astronautics International Conference on Low-Cost Planetary Missions.

8. Gene Therapy

Walters, Leroy. 1986. The ethics of human gene therapy. *Nature* 320 (March 20): 225–227.

Anderson, W. F. 1992. Human gene therapy. *Science* 256 (May 8): 808–813.

Culver, Kenneth W. 1994. *Gene therapy: A handbook for physicians.* Mary Ann Liebert, Inc., Publishers.

Lyon, Jeff, and Peter Gorner. 1995. *Altered fates: Gene therapy and the retooling of human life.* W. W. Norton, 1995.

9. Artificial Organs

Langer, Robert. 1990. New methods of drug delivery. *Science* 249 (September 28): 1527–1533.

Langer, Robert, and Joseph P. Vacanti. 1993. Tissue engineering. *Science* 260 (May 14): 920–926.

Mooney, D. J., G. Organ, Joseph P. Vacanti and Robert Langer. 1994. Design and fabrication of biodegradable polymer devices to engineer tubular tissues. *Cell Transplantation* 3:203–210.

Roush, Wade. 1995. Envisioning an artificial retina. *Science* 268 (May 5): 637–638.

10. Future Contraceptives

Leyton, L., et al. 1992. Regulation of mouse gamete interaction by sperm tyrosine kinase. *Proceedings of the National Academy of Sciences USA* 89 (December 15): 11692–11695.

Benoff, S., et al. 1993. Fertilization potential in vitro is correlated with head-specific mannose-ligand receptor expression, acrosome status and membrane cholesterol content. *Human Reproduction* 8 (December 1): 2155–2166.

Alexander, N. J., and G. Bialy. 1994. Contraceptive vaccine development. *Reproduction Fertility and Development* 6:273–280.

Dym, M. 1994. Spermatogonial stem cells of the testis. *Proceedings of the National Academy of Sciences USA* 91 (November 22): 11287–11289.

11. Self-Assembling Materials

Fendler, Janos H. 1982. *Membrane mimetic chemistry.* John Wiley & Sons.

Whitesides, George M., J. P. Mathias and C. T. Seto. 1991. Molecular self-assembly and nanochemistry: A chemical strategy for the synthesis of nanostructures. *Science* 254 (November 29): 1312–1319.

Ball, Philip. 1994. *Designing the molecular world: Chemistry at the frontier.* Princeton University Press.

Singhvi, R., A. Kumar, G. P. Lopez, G. N. Stephanopoulos, D. I. C. Wang, G. M. Whitesides and D. E. Ingber. Engineering cell shape and function. *Science* 264 (April 29): 696–698.

12. Engineering Microscopic Machines

Petersen, Kurt. 1982. Silicon as a mechanical material. *Proceedings of the IEEE* 70 (May): 420–457.

Angell, James B., Stephen C. Terry and Phillip W. Barth. 1983. Silicon micromechanical devices. *Scientific American* 248 (April): 44–55.

Core, Theresa A., et al. 1993. Fabrication technology for an integrated surface-micromachined sensor. *Solid-State Technology* 36 (October): 39–47.

13. Intelligent Materials

Amato, Ivan. 1992. Animating the material world. Science *255 (January 17): 284–286.*

Dry, Carolyn M. 1993. Passive smart materials for sensing and actuation. *Journal of Intelligent*

*Materials Systems and Structures* 4 (July): 415.

Chorpa, Inderjit, ed. 1995. *Smart structures and materials 1995: Smart structures and integrated systems.* International Society for Optical Engineering, Bellingham, Wash., 1995.

14. High-Temperature Superconductors

Müller, K. Alex, and J. Georg Bednorz. 1987. The discovery of a class of high-temperature superconductors. *Science* 237 (September 4): 1133–1139.

Chu, C. W. 1987. Superconductivity above 90K. *Proceedings of the National Academy of Sciences USA* 84 (July): 4681–4682.

Hazen, Robert M. 1989. *The breakthrough: The race for the superconductor.* Summit Books, 1989.

Orlando, Terry P., and Kevin A. Delin. 1991. *Foundations of applied superconductivity.* Addison-Wesley.

Sheahen, T. P. 1994. *Introduction to high temperature superconductivity.* Plenum Press.

Ginsberg, Donald M. 1989, 1990, 1992 and 1994. *Physical properties of high-temperature superconductors,* vols. 1-4. World Scientific, Singapore and Teaneck, N.J.

15. Solar Energy

Zweibel, Kenneth, Paul Hersch and Solar Energy Research Institute. 1984. *Basic photovoltaic principles and methods.* Van Nostrand Reinhold.

Gordon, Deborah, and Union of Concerned Scientists. 1991. *Steering a new course: Transportation, energy and the environment.* Island Press.

Johansson, Thomas B., Henry Kelly, Amulya K. N. Reddy, Robert Williams and Laurie Burnham. 1993. *Renewable energy: Sources for fuels and electricity.* Island Press.

Hubbard, Harold M., Paul Notari, Satyen Deb and Shimon Awerbach. 1994. *Progress in solar energy technologies and applications.* American Solar Energy Society.

16. Fusion

Conn, Robert W. 1983. The engineering of magnetic fusion reactors. *Scientific American* 249 (October): 60–71.

Craxton, R. Stephen, Robert L. McCrory and John M. Soures. 1986. Progress in laser fusion. *Scientific American* 254 (August): 68–79.

Herman, Robin. 1990. *Fusion: The search for endless energy.* Cambridge University Press.

17. The Industrial Ecology of the 21st Century

Frosch, Robert A., and Nicholas E. Gallopoulos. 1989. Strategies for manufacturing. *Scientific American* 261 (September): 144–152.

Allenby, Braden R., and Deanna J. Richards, eds. 1994. *The greening of industrial ecosystems.* National Academy Press.

Frosch, Robert A. 1994. Industrial ecology: Minimizing the impact of industrial waste. *Physics Today* 47 (November): 63–68.

Ayres, R., and U. Simonis, eds. 1994. *Industrial metabolism: Restructuring for sustainable development.* United Nations University Press.

18. Technology for a Sustainable Agriculture

Cohen, Joel I., J. Trevor Williams, Donald L. Plucknett and Henry Shands. 1991. Ex situ conservation of plant genetic resources: Global development and environmental concerns. *Science* 253 (August 23): 866–872.

Gasser, Charles S., and Robert T. Fraley. 1992. Transgenic crops. *Scientific American* 266 (June): 62–69.

Wilson, Edward O. 1993. *The diversity of life.* W. W. Norton.

Waggoner, Paul E. 1994. *How much land can ten billion people spare for nature?* Council for Agricultural Science and Technology Report No. 121.

# Sources of the Photographs

Charles O'Rear: Figures 1.1, 1.2 and "The Limits of Lithography"
Henryk Temkin, AT&T Bell Laboratories: "And After 2020?"

Richard Pasley: Figure 2.1 and "Wireless Telephony for Developing Countries"

Jason Goltz: Figure 3.1 (*left*)
Steven E. Sutton, Dumo: Figure 3.1 (*right*)

Tony Stone Images: Figure 4.1

Michael Rosenfeld, Tony Stone Images: Figure 5.1
Mercedes-Benz: Figure 5.2

Remi Benali, Gamma Liaison: Figure 6.1 "Pilots"
Alan H. Epstein: Figure 6.1 "Smart engines"
Pratt & Whitney: Figure 6.1 "High-bypass jet engine"
David Scharf: Figure 6.1 "Microactuators"
Textron Specialty Materials: Figure 6.2

Z. Aronovsky, Zuma: Figure 7.1
National Aeronautics and Space Administration: Figure 7.3

Alex S. MacLean, Landslides: "Commentary: Why Go Anywhere?"

Jessica Boyatt: Figure 8.1

Cygnus Therapeutic Systems: Figure 9.1
David J. Mooney: Figure 9.2

Nancy J. Alexander: Figure 10.1

Ralph L. Brinster, University of Pennsylvania School of Veterinary Medicine: "Tomorrow's Infertility Treatments"

George M. Whitesides: Figures 11.2 and 11.3
Yuan Lu, Harvard University: Figure 11.4
Christoph Burki, Tony Stone Images: "Two Types of Self-Assembly" (raindrops)
Neil Harding, Tony Stone Images: "Two Types of Self-Assembly" (embryo)

Analog Devices: Figure 12.2
Cornell University: Figure 12.3

Etrema Products, Inc.: Figure 13.1
AP/Wide World Photos: "Advanced Composites"
Bob Sacha: "Custom Manufacturing"

Texas Center for Superconductivity, University of Houston: Figure 14.1
American Superconductor: Figure 14.3

Barrie Rokeach: Figure 15.1 (*background*)
Alex S. McLean, Landslides: Figures 15.1a, b, c and d

Roger Ressmeyer, Starlight: Figures 16.1 and "Disposing of Nuclear Waste"
Princeton Plasma Physics Laboratory: Figure 16.2

Louis J. Circeo, Georgia Institute of Technology: Chapter 17, "The Ultimate Incinerators"

Patrick Aventurier, Gamma Liaison: Chapter 18, "The Next Wave: Aquaculture"

# INDEX

Page numbers in *italics* indicate illustrations.